Lecture Notes in Electrical Engineering

Volume 809

The book series *Lecture Notes in Electrical Engineering* (LNEE) publishes the latest developments in Electrical Engineering - quickly, informally and in high quality. While original research reported in proceedings and monographs has traditionally formed the core of LNEE, we also encourage authors to submit books devoted to supporting student education and professional training in the various fields and applications areas of electrical engineering. The series cover classical and emerging topics concerning:

- Communication Engineering, Information Theory and Networks
- Electronics Engineering and Microelectronics
- Signal, Image and Speech Processing
- Wireless and Mobile Communication
- Circuits and Systems
- Energy Systems, Power Electronics and Electrical Machines
- Electro-optical Engineering
- Instrumentation Engineering
- Avionics Engineering
- Control Systems
- Internet-of-Things and Cybersecurity
- Biomedical Devices, MEMS and NEMS

For general information about this book series, comments or suggestions, please contact leontina.dicecco@springer.com.

To submit a proposal or request further information, please contact the Publishing Editor in your country:

China

Jasmine Dou, Editor (jasmine.dou@springer.com)

India, Japan, Rest of Asia

Swati Meherishi, Editorial Director (Swati.Meherishi@springer.com)

Southeast Asia, Australia, New Zealand

Ramesh Nath Premnath, Editor (ramesh.premnath@springernature.com)

USA, Canada:

Michael Luby, Senior Editor (michael.luby@springer.com)

All other Countries:

Leontina Di Cecco, Senior Editor (leontina.dicecco@springer.com)

**** This series is indexed by EI Compendex and Scopus databases. ****

More information about this series at https://link.springer.com/bookseries/7818

Alfonso Bachiller Soler · Ramón Cano Gonzalez ·
Miguel Angel González Cagigal

Solved Problems for Transient Electrical Circuits

Springer

Alfonso Bachiller Soler
Department of Electrical Engineering
University of Seville
Seville, Spain

Ramón Cano Gonzalez
Department of Electrical Engineering
University of Seville
Seville, Spain

Miguel Angel González Cagigal
Department of Electrical Engineering
University of Seville
Seville, Spain

ISSN 1876-1100 ISSN 1876-1119 (electronic)
Lecture Notes in Electrical Engineering
ISBN 978-3-030-88143-6 ISBN 978-3-030-88144-3 (eBook)
https://doi.org/10.1007/978-3-030-88144-3

This Springer imprint is published by the registered company Springer Nature Switzerland AG
The registered company address is: Gewerbestrasse 11, 6330 Cham, Switzerland

To our parents

Foreword

The title of this book, *Solved Problems for Transient Electrical Circuits*, clearly indicates that its contents deal with one of the fundamental themes of Electrical Circuit Theory, as does that of "Three-Phase Circuits" and "Dependent Sources", in the context of Electrical Engineering. However, in my opinion, the importance of transient circuit behaviour surpasses the limits of electrical engineering, since it also constitutes a fundamental component of electronic engineering, mainly in its digital, power, and telecommunications areas. The study of the circuits under a dynamic regime enables the correct interpretation to be attained of certain types of electrical behaviours that remain elusive, even to engineers, in a first reasoning. Examples of such behaviours include the emergence of voltages far higher than those of the generators themselves in circuits and in electrical networks, with destructive effects; the untimely firing of differential switches in homes; and the sporadic performance of protection during transformer commissioning.

Transients are also present in digital electronic circuits, since their binary nature obliges transistors to work as ideal switches at high speeds, thereby generating a transient period in each switching. The same is true of power electronics, where, for performance reasons, semiconductors also work as switches by switching thousands of amps into microsecond fractions. In these cases, the permanent regime of the circuits becomes a continuous sequence of transient regimes.

Transient operation in electronic devices involves rapid variations in voltage and current that can cause electromagnetic disturbances in the circuits. Electromagnetic Compatibility regulatory requirements constitute one of the key points of electronic design and have emerged as a discipline in telecommunications and industrial engineering studies.

The importance of the study of the dynamic regime of circuits, both electrical and electronic, should be borne in mind for future professionals of these subjects.

The study of transients is addressed in this book by first stating the fundamental theoretical concepts, which are subsequently consolidated with the help of solved and annotated problems of increasing difficulty. Although, in professional practice, complex circuits are often solved by simulation and not by mathematical tools such as differential equations and the Laplace transform, the interpretation and valuation of numerical results remains essential, and this is only possible if the necessary theoretical knowledge is available.

The text is organised into three chapters: First-Order Transients, Second-Order Transients, and the Laplace Transform. Each chapter begins with the corresponding theory, supported by application examples, followed by a collection of resolved and annotated problems. In all three parts, a suitable balance is struck between the theory necessary to understand the concepts and their application through problem-solving. The authors, as in all their publications, have managed to provide the reader with a book containing figures and typography both in harmony and of the highest quality.

Finally, with the help of this book, the reader, as a student, is in possession of the tools necessary to successfully master the subjects related to electrical circuits, and, as a professional, can preserve this volume in order to extract the fundamentals of its content.

Sevilla, Spain
June 2021

Vicente Simón Sempere

Preface

The text is intended for the first course on electric circuits. The focus is on the transient response of linear circuits. The analysis of this type of circuits is generally carried out in the second year of electrical engineering studies and related fields.

The book has been divided into three large chapters that progressively address the study of the transient response of first-order and second-order circuits and, finally, circuits of any order through the use of the Laplace transform.

Each block begins with a detailed study of the theoretical knowledge and the resolution techniques necessary to obtain the transient response of the different types of circuits. This is followed by a significant number of solved problems. The resolution of each exercise has been carried out in detail and with the support of more than 300 figures. For a better understanding of the transitory phenomenon, the evolution of the voltages and currents of the elements during the transient period has been graphically represented in the cases considered relevant. The exercises have been ordered from the most elementary to the most complex, which allows progressive learning.

In this book, only circuits with linear elements are considered where the origin of the transient is caused by the opening or closing of switches. This model responds to maneuvering operations in high and low voltage electrical networks. It is also the model that is analysed when considering the main defects in this type of installations, such as short circuits and insulation defects. Finally, the same model is obtained in power electronics topics, where electronic devices (BJT, MosFet,...) have the behaviour of switches that switch several times per second.

Although several circuits have been included that contain controlled sources, which may lead to instability, the parameters have been selected to obtained stable circuits in all cases. The analysis of unstable circuits is carried out initially with the same techniques, but it leads to non-real results if the non-linearity of the elements is not considered.

In the problems, voltages and currents of the elements have been the variables under study. Power and energy can be easily obtained from them. However, it has been preferred not to include them in the calculations to obtain a more fluent reading of the text.

Sevilla, Spain
June 2021

Alfonso Bachiller Soler
Ramón Cano González
Miguel Angel González Cagigal

Contents

1 First Order Transients ... 1
1.1 Introduction ... 1
1.2 First Order Circuits ... 2
1.2.1 RC Circuits ... 2
1.2.2 RL Circuits ... 5
1.2.3 Generic Differential Equation of a First Order Circuit ... 6
1.3 Transient Response of First Order Circuits ... 7
1.3.1 Natural Response ... 8
1.3.2 Forced or Steady State Response ... 9
1.3.3 Complete Response ... 9
1.3.4 Initial Conditions ... 10
1.4 Generalization of the Transient Response ... 11
1.5 Procedure to Obtain the Response of a First Order Circuit ... 13
1.6 Impulse Response ... 17
1.6.1 Parallel Connection of Capacitors ... 18
1.6.2 Series Connection of Inductors ... 19
1.7 Solved Problems ... 21

2 Second Order Transients ... 97
2.1 Second Order Circuits ... 97
2.1.1 Series RLC Circuit ... 98
2.1.2 Parallel RLC Circuit ... 101
2.1.3 Generic Differential Equation of a Second Order Circuit ... 103
2.2 Transient Response of Second Order Circuits ... 105
2.2.1 Natural Response ... 105
2.2.2 Forced or Steady State Response ... 107
2.2.3 Complete Response ... 107
2.2.4 Initial Conditions ... 108
2.3 Procedure to Obtain the Response of a Second Order Circuit ... 110
2.4 Solved Problems ... 113

3 Laplace Transform Analysis . . . 171
3.1 Introduction . . . 171
3.2 Definition . . . 171
3.3 Main Properties and Theorems . . . 172
3.4 Laplace Transform Pairs . . . 173
3.5 Application to the Analysis of Electrical Circuits . . . 173
3.5.1 Introduction . . . 173
3.5.2 Voltage-Current Relationship in the s-Domain . . . 174
3.5.3 Impedance and Admittance . . . 180
3.5.4 Kirchhoff's Laws . . . 180
3.5.5 Methodology of Resolution . . . 181
3.6 Inverse Laplace Transform . . . 183
3.6.1 Inverse Laplace Transform Calculation Methodology . . . 184
3.6.2 Simple Real Poles, $p_1 \neq p_2 \neq \cdots \neq p_m$ . . . 185
3.6.3 Multiple Real Pole . . . 186
3.6.4 Complex-Conjugated Pole . . . 187
3.7 Solved Problems . . . 188

Chapter 1
First Order Transients

Abstract This chapter covers the first order circuits, beginning with a theoretical introduction of the concepts required to correctly address each of the subsequent problems. A total of 31 fully solved problems with explanatory comments are included.

1.1 Introduction

In circuits with resistors as sole passive elements, voltages and currents respond immediately to the changes in the sources of excitation. In such circuits, known as static circuits, the voltage and current values of the involved elements are given by a set of algebraic equations, meaning that each instant of time can be independently analyzed. This is not the case in circuits containing energy storage elements, i.e. inductors or capacitors, where the voltage is related to the current through a differential equation, resulting in a dynamic response of the circuit. In this type of circuits (dynamic circuits), information on the past is necessary to determine the response at any time.

In dynamic circuits excited with dc or ac sources, after a period of time has elapsed (transient regime) the so-called steady-state regime is reached, where the response is stabilized at a constant value (dc excitation) or a periodic wave (ac excitation). As an illustrative example, the response of an RC circuit is represented in Figs. 1.1 and 1.2, including respectively a dc voltage source and an ac voltage source. It can be seen how, after the transient period, the steady-state regime is reached in both cases.

In general terms, the transition from a steady-state regime to a different one involves a transient period. The origin of these transient processes can be related to various actions, including opening and closing of switches, short circuits or any other variation in the circuit topology or parameters.

A. Bachiller Soler et al., *Solved Problems for Transient Electrical Circuits*,
Lecture Notes in Electrical Engineering 809,
https://doi.org/10.1007/978-3-030-88144-3_1

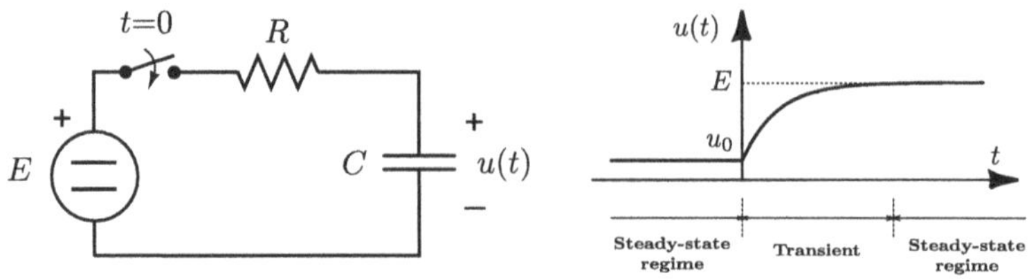

Fig. 1.1 Connection of an RC circuit to a dc source

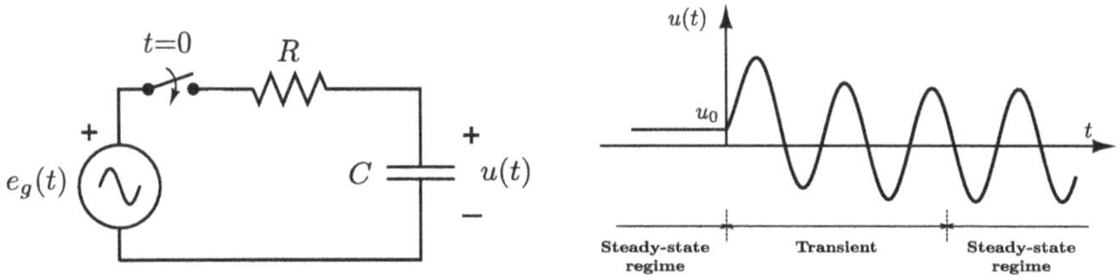

Fig. 1.2 Connection of an RC circuit to an ac source

1.2 First Order Circuits

First order circuits are defined as those where any voltage or current can be obtained using a first order differential equation. Some examples of first order circuits are:

- Circuits with a single electrical energy storage element: inductor or capacitor, Fig. 1.3.

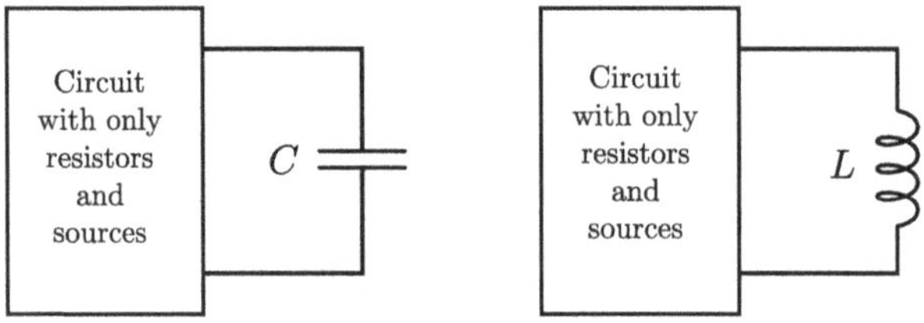

Fig. 1.3 First order circuits with one energy storage element

- Circuits including multiple energy storage elements of the same type, which can be combined into a single equivalent element, Figs. 1.4 and 1.5.

1.2.1 RC Circuits

First order circuits will be considered with one or more capacitors that can be combined into a single equivalent one. The rest of the circuit, composed by electrical

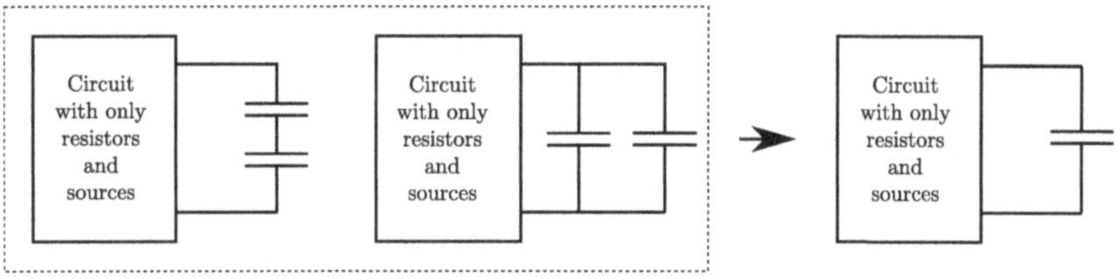

Fig. 1.4 Circuit with two capacitors connected in series and in parallel

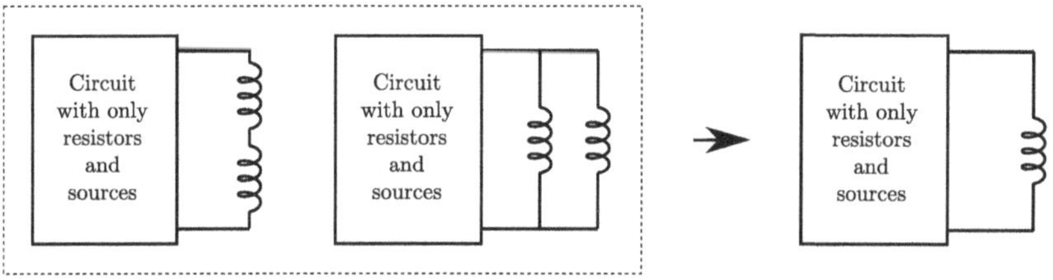

Fig. 1.5 Circuit with two inductors connected in series and in parallel

sources and resistors, can be replaced by its Thévenin equivalent as shown in Fig. 1.6. This way, the study of the RC series circuit excited by a voltage source encompasses all the first order circuits whose storage element is a capacitor.

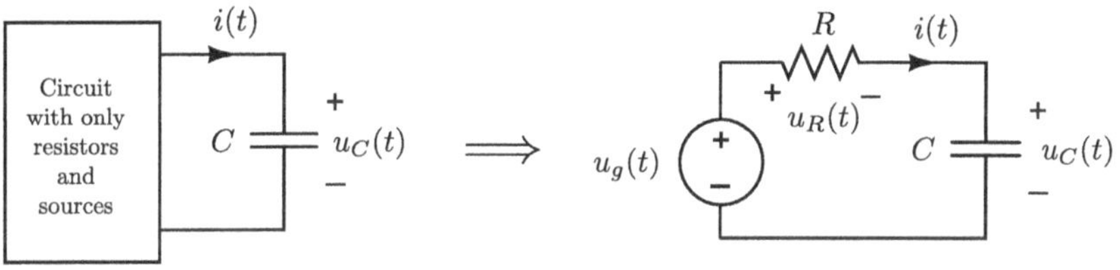

Fig. 1.6 RC circuit and its Thévenin equivalent

The differential equations that defines the behavior of the variables involved in the circuit represented in Fig. 1.6 will be subsequently obtained.

Differential equation of capacitor voltage

Applying Kirchhoff's voltage law:

$$u_C(t) + u_R(t) = u_g(t) \tag{1.1}$$

The resistor voltage can be substituted using Ohm's law:

$$u_C(t) + R \cdot i(t) = u_g(t) \tag{1.2}$$

The capacitor defining equation is now considered

$$i(t) = C\frac{du_C(t)}{dt} \tag{1.3}$$

and after rearranging terms, the final equation yields:

$$\frac{du_C(t)}{dt} + \frac{1}{RC}u_C(t) = \frac{u_g(t)}{RC} \tag{1.4}$$

Differential equation of current

Differentiating Eq. (1.2) gives:

$$\frac{du_C(t)}{dt} + R\frac{di(t)}{dt} = \frac{du_g(t)}{dt} \tag{1.5}$$

The defining equation of the capacitor (1.3) is considered again,

$$\frac{i(t)}{C} + R\frac{di(t)}{dt} = \frac{du_g(t)}{dt} \tag{1.6}$$

and rearranging terms:

$$\frac{di(t)}{dt} + \frac{1}{RC}i(t) = \frac{1}{R}\frac{du_g(t)}{dt} \tag{1.7}$$

Differential equation of the resistor voltage

If Eq. (1.1) is differentiated, the following expression is obtained:

$$\frac{du_C(t)}{dt} + \frac{du_R(t)}{dt} = \frac{du_g(t)}{dt} \tag{1.8}$$

and the capacitor defining equation (1.3) is used, yielding:

$$\frac{i(t)}{C} + \frac{du_R(t)}{dt} = \frac{du_g(t)}{dt} \tag{1.9}$$

Finally, applying Ohm's law on the resistor and rearranging terms:

$$\frac{du_R(t)}{dt} + \frac{1}{RC}u_R(t) = \frac{du_g(t)}{dt} \tag{1.10}$$

It can be verified from the differential equations obtained for each variable, (1.4), (1.7) and (1.10), that all of them have the same coefficients and differ only in the independent term.

1.2.2 RL Circuits

First order circuits with one inductor or a group of them that can be combined into a single equivalent are now considered. The rest of the circuit is exclusively made up of electrical sources and resistors, without energy storage elements, so that it can be replaced by its Norton equivalent, which consists of a current source in parallel with a resistor, as shown in Fig. 1.7.

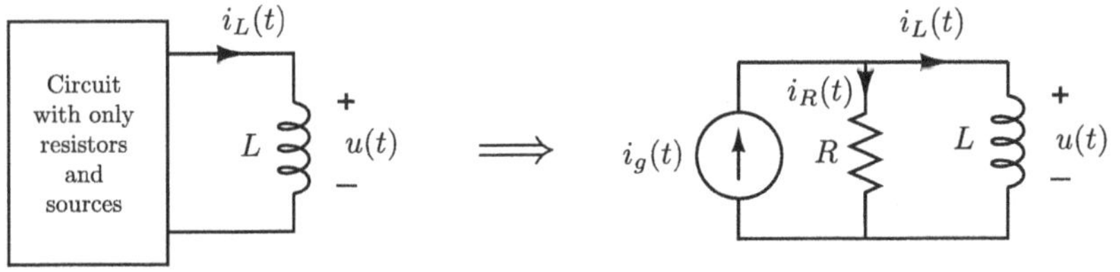

Fig. 1.7 RL circuit and its Norton equivalent

The differential equations that defines the behavior of the variables involved in the parallel RL circuit represented in Fig. 1.7 will be subsequently obtained.

Inductor current equation

Applying Kirchhoff's current law:

$$i_L(t) + i_R(t) = i_g(t) \tag{1.11}$$

The resistor current is substituted using Ohm's law

$$i_L(t) + \frac{u(t)}{R} = i_g(t) \tag{1.12}$$

and considering the inductor defining equation

$$u(t) = L\frac{di_L(t)}{dt} \tag{1.13}$$

the final expression yields, after rearranging terms:

$$\frac{di_L(t)}{dt} + \frac{R}{L}i_L(t) = \frac{R}{L}i_g(t) \tag{1.14}$$

Circuit voltage equation

Differentiating Eq. (1.12) gives:

$$\frac{di_L(t)}{dt} + \frac{1}{R}\frac{du(t)}{dt} = \frac{di_g(t)}{dt} \tag{1.15}$$

Using the inductor defining equation (1.13)

$$\frac{u(t)}{L} + \frac{1}{R}\frac{du(t)}{dt} = \frac{di_g(t)}{dt} \tag{1.16}$$

and rearranging terms:

$$\frac{du(t)}{dt} + \frac{R}{L}u(t) = R\frac{di_g(t)}{dt} \tag{1.17}$$

Equation of resistor current

Equation (1.11) is differentiated

$$\frac{di_L(t)}{dt} + \frac{di_R(t)}{dt} = \frac{di_g(t)}{dt} \tag{1.18}$$

and the inductor defining equation (1.13) is used as follows:

$$\frac{u(t)}{L} + \frac{di_R(t)}{dt} = \frac{di_g(t)}{dt} \tag{1.19}$$

Finally, Ohm's law is applied on the resistor and the terms are rearranged, resulting:

$$\frac{di_R(t)}{dt} + \frac{R}{L}i_R(t) = \frac{di_g(t)}{dt} \tag{1.20}$$

It can be noticed that the coefficients of the previous differential equations, (1.14), (1.17) and (1.20), are the same in all cases, only differing in the independent terms. This fact allows the derivation of a generic equation for all the variables.

1.2.3 Generic Differential Equation of a First Order Circuit

Figure 1.8 shows the differential equations of all the variables involved in the RC and RL circuits, as obtained in the previous sections.

According to Fig. 1.8, it is observed how all equations can be expressed in the following generic form:

$$\frac{df(t)}{dt} + \frac{1}{\tau}f(t) = g(t) \tag{1.21}$$

where $f(t)$ denotes the considered voltage or current, $g(t)$ is a function related to the excitation source of the circuit and τ is known as the circuit time constant, whose SI unit is the second, and depends on the parameter values for the passive elements of the circuit. This constant is characteristic of each circuit and its value is given by:

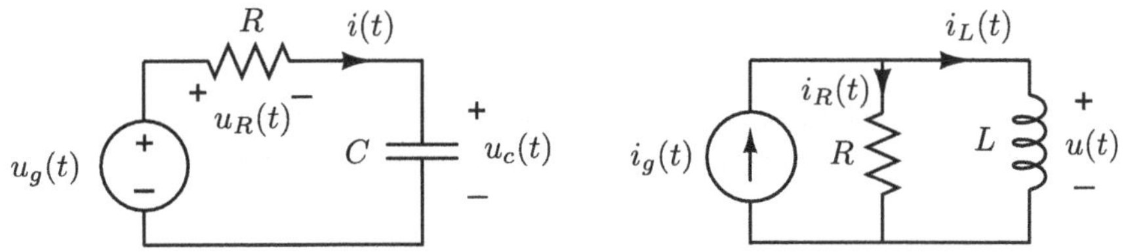

$$\frac{du_c(t)}{dt} + \frac{1}{RC}u_c(t) = \frac{u_g(t)}{RC} \qquad \frac{di_L(t)}{dt} + \frac{R}{L}i_L(t) = \frac{R}{L}i_g(t)$$

$$\frac{di(t)}{dt} + \frac{1}{RC}i(t) = \frac{1}{R}\frac{du_g(t)}{dt} \qquad \frac{du(t)}{dt} + \frac{R}{L}u(t) = R\frac{di_g(t)}{dt}$$

$$\frac{du_R(t)}{dt} + \frac{1}{RC}u_R(t) = \frac{du_g(t)}{dt} \qquad \frac{di_R(t)}{dt} + \frac{R}{L}i_R(t) = \frac{di_g(t)}{dt}$$

Fig. 1.8 Generic differential equations of RC and RL circuits

- RC circuit: $\tau = R \cdot C$
- RL circuit: $\tau = L/R$

It should be remarked that, if there are several capacitors in the first order circuit which can be grouped into one, the value of C represents the equivalent capacity. Similarly, the value of L represents the coefficient of self-induction of the equivalent inductor. Finally, R is the equivalent resistance of the passive circuit seen from the L or C terminals. Therefore, the most general expressions for the constant τ are:

- RC circuit: $\tau = R_{eq} \cdot C_{eq}$
- RL circuit: $\tau = L_{eq}/R_{eq}$

1.3 Transient Response of First Order Circuits

As explained before, all the voltages and currents of the first order circuits are given by the following constant-coefficient linear differential equation:

$$\frac{df(t)}{dt} + \frac{1}{\tau}f(t) = g(t) \tag{1.22}$$

The solution of this equation is used to obtain the transient response of the different variables of the circuit under study. Mathematically, the general solution of this type of equations can be expressed as the sum of the solution of the homogeneous equation plus a particular solution of the complete equation. In electrical circuits, the first term is known as the natural response of the circuit, while the particular solution is known as the forced or steady-state response. Therefore, the solution of the differential equation (1.22) can be expressed as

$$f(t) = f_n(t) + f_p(t) \tag{1.23}$$

where $f_n(t)$ is the natural response of the circuit and $f_p(t)$ represents the forced or steady-state response. The derivation of each term is described below.

1.3.1 Natural Response

The natural response corresponds, mathematically, to the solution of the homogeneous differential equation, that is, the independent term equal to zero

$$\frac{df(t)}{dt} + \frac{1}{\tau} f(t) = 0 \tag{1.24}$$

whose solution for $t \geq 0$ is:

$$f_n(t) = K \cdot \mathrm{e}^{-t/\tau} \tag{1.25}$$

It should be remarked that, as in the homogeneous Eq. (1.24) the term $g(t)$ (related to the excitation source) does not appear, the solution of this equation corresponds to the circuit response if the excitation sources were canceled, hence it is called the natural response of the circuit.

The natural response of a first order circuit has an exponential nature, being the rate of decrease determined by the value of the time constant, τ. It can be observed in Fig. 1.9 that, after a time period equal to the value of the time constant has elapsed, the natural response has decreased from its initial value, K, to $0{,}368K$, that is, it has been reduced a 63,2 %. Although mathematically the natural response never disappears, in practice it can be considered that, after a time period equal to 5τ, the natural response is negligible, since its value has been reduced to only $0{,}007K$.

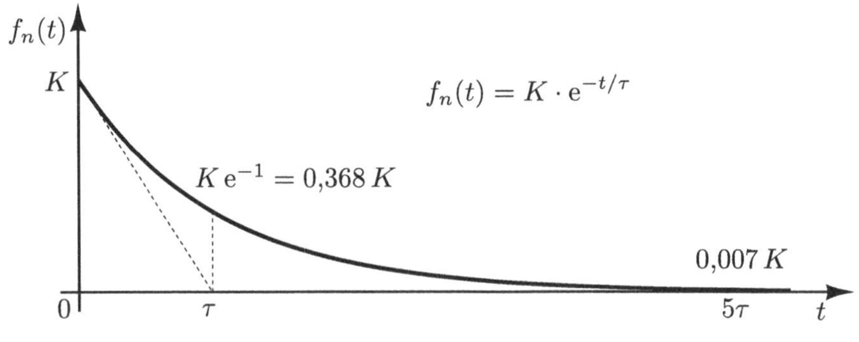

Fig. 1.9 Evolution of the natural response

Finally, it can be noticed from Fig. 1.10 that a higher value of the time constant corresponds to a longer duration of the natural response, as expected.

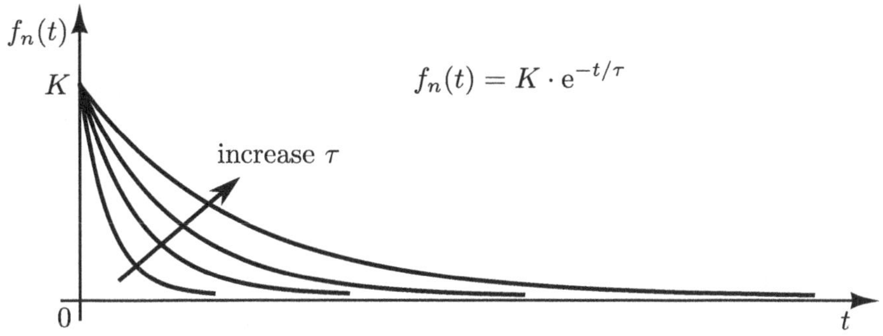

Fig. 1.10 Influence of the time constant on the natural response

1.3.2 Forced or Steady State Response

The forced response corresponds mathematically to a particular solution of the complete differential equation. The nature of this particular solution is usually the same as that of the independent term $g(t)$, that is, the same type of response as the excitation of the circuit. Since the natural response vanishes, the forced response is the only one remaining in time, hence in electrical circuits it is also known as the steady-state response.

Several methods can be used to obtain a particular solution of a constant-coefficient differential equation, such as parameter variation or undetermined coefficients, among others. However, in the particular case of electrical circuits under dc or ac supply, specific techniques have been developed to obtain the permanent regime. Therefore, the forced response will be obtained using these techniques.

1.3.3 Complete Response

Once the natural and the steady-state responses are known, the complete response of the circuit under study will be:

$$f(t) = f_n(t) + f_p(t) = K \cdot e^{-t/\tau} + f_p(t) \tag{1.26}$$

When the steady-state response, $f_p(t)$, and the time constant of the circuit, τ, have been obtained, the last step is the calculation of the constant K, in order to fully characterize the response of the considered variable.

The value of this constant is obtained from the initial value of the variable ($f(0^+)$). Thus, in $t = 0^+$ it must be verified:

$$f(0^+) = K + f_p(0^+) \tag{1.27}$$

from where it can be derived

$$K = f(0^+) - f_p(0^+) \tag{1.28}$$

yielding the final expression:

$$f(t) = f_p(t) + \left[f(0^+) - f_p(0^+)\right] \cdot \mathrm{e}^{-t/\tau} \tag{1.29}$$

This expression allows the calculation of the voltage or current of any element in a first order circuit, where:

- $f(t)$ is the considered voltage or current.
- $f_p(t)$ is the steady-state response of this variable.
- $f_p(0^+)$ is the value in $t = 0^+$ of the steady-state response.
- τ is the time constant of the circuit.
- $f(0^+)$ is the initial value of the mentioned voltage or current.

The calculation of $f(0^+)$ in first order circuits will be described in the next section.

1.3.4 Initial Conditions

The transition from a steady-state regime to a different one is generally determined by a transient period involving a redistribution of the energy stored in inductors and capacitors. Additionally, a variation in the energy state of the electrical sources is also produced. Since an instantaneous energy redistribution is not possible, the following points must be verified in absence of impulse responses:

- The capacitor voltage cannot suffer discontinuities:

$$u_C(0^+) = u_C(0^-)$$

- The inductor current cannot suffer discontinuities:

$$i_L(0^+) = i_L(0^-)$$

With these premises, the initial value of any voltage or current, $f(0^+)$, can be calculated by solving a circuit where:

1. The excitation sources, $e_g(t)$ and $i_g(t)$, are replaced by two sources of constant value:

$$E_g = e_g(0^+) \ ; \ \ I_g = i_g(0^+)$$

2. In the case of an RC circuit, the capacitor is replaced by a constant voltage source, whose value is given by:

$$u_C(0^+) = u_C(0^-) = U_0$$

3. In RL circuits, the corresponding inductor is replaced by a constant current source, whose value is given by:

$$i_L(0^+) = i_L(0^-) = I_0$$

Figures 1.11 and 1.12 summarize the procedure for first order circuits with a capacitor and an inductor, respectively.

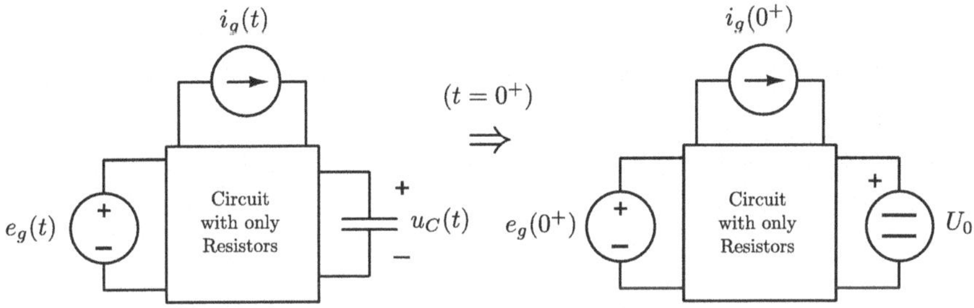

Fig. 1.11 Derivation of the circuit at $t = 0^+$. Circuit with capacitor

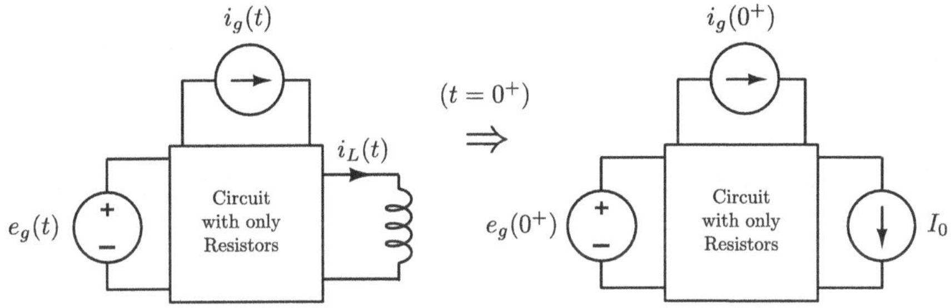

Fig. 1.12 Derivation of the circuit at $t = 0^+$. Circuit with inductor

1.4 Generalization of the Transient Response

In the previous sections it was assumed that the transient period under analysis begins at instant $t = 0$. However, the obtained results can be generalized for any time $t = t_0$. This fact will be applied to the analysis of several concatenated transient periods. For time-invariant systems, if the transient period started at $t = t_0$, the complete response would be given by the following expression:

$$f(t) = f_p(t) + \left[f(t_0^+) - f_p(t_0^+)\right] e^{-(t-t_0)/\tau} \quad (1.30)$$

As mentioned before, for the calculation of the initial conditions, and in the absence of impulse responses, it must be verified that:

- The voltage capacitor cannot suffer discontinuities:

$$u_C(t_0^+) = u_C(t_0^-) \tag{1.31}$$

- The current inductor cannot suffer discontinuities:

$$i_L(t_0^+) = i_L(t_0^-) \tag{1.32}$$

With these premises, the initial value of any voltage or current, $f(t_0^+)$, can be calculated by solving a circuit where:

1. The excitation sources, $e_g(t)$ and $i_g(t)$, are replaced by two sources of constant value:
$$E_g = e_g(t_0^+) \;\; ; \;\; I_g = i_g(t_0^+)$$
2. In the case of an RC circuit, the capacitor is replaced by a constant voltage source, whose value is given by:
$$u_C(t_0^+) = u_C(t_0^-) = U_{t_0}$$
3. In RL circuits, the corresponding inductor is replaced by a constant current source, whose value is given by:
$$i_L(t_0^+) = i_L(t_0^-) = I_{t_0}$$

Figures 1.13 and 1.14 summarize the described procedure to obtain the circuit at $t = t_0^+$, from which the initial value of any variable can be obtained at the considered instant.

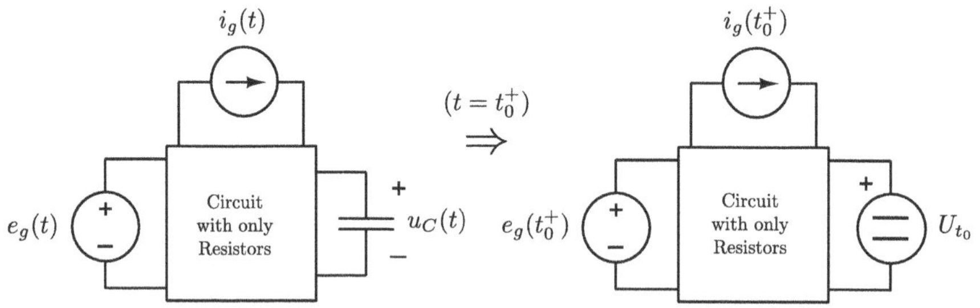

Fig. 1.13 Derivation of the circuit at $t = t_0^+$. Circuit with capacitor

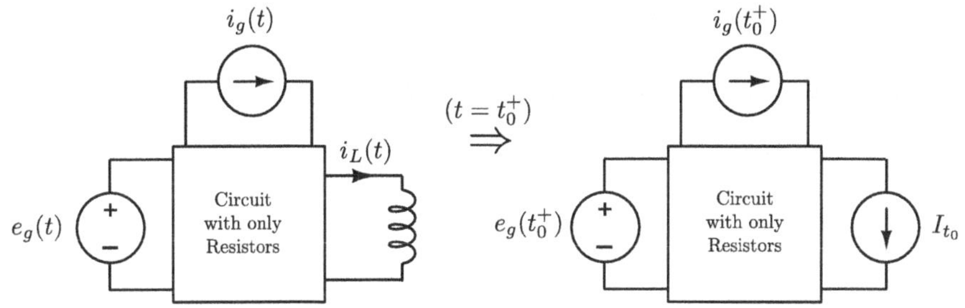

Fig. 1.14 Derivation of the circuit at $t = t_0^+$. Circuit with inductor

1.5 Procedure to Obtain the Response of a First Order Circuit

In essence, the steps to follow in order to obtain the voltage or current of any element involved in a first order circuit during a transient period starting at t_0^+ are listed below. Let be $f(t)$ the variable to be calculated, its time evolution is given by:

$$f(t) = f_p(t) + \left[f(t_0^+) - f_p(t_0^+)\right] \cdot \mathrm{e}^{-(t-t_0)/\tau} \tag{1.33}$$

Each term in the previous expression is calculated as follows:

1. For the derivation of the steady-state response of the interest variable, $f_p(t)$ (with $t \geq t_0$), any of the known techniques for steady-state analysis of circuits under dc or ac supply can be used, depending on the electrical sources involved. The value of $f_p(t_0^+)$ is calculated making $t = t_0$ in the obtained expression for $f_p(t)$. In the case that the circuit for $t \geq t_0$ lacks independent sources of excitation, $f_p(t)$ will be null.
2. The value of the time constant τ is given by $\tau = R_{eq}C_{eq}$ (resp. $\tau = L_{eq}/R_{eq}$) for $t \geq t_0$, if the circuit has capacitors (resp. inductors). First, the passive circuit is obtained by canceling the independent sources, then all the capacitors (resp. inductors) are combined into an equivalent one. If this combination is not possible, the circuit under study is not a first order one. Finally, the equivalent resistance of the circuit is determined from the terminals of the equivalent capacitor (resp. inductor).
3. The initial value of the variable considered, $f(t_0^+)$, is determined using the voltage of the capacitor $u_C(t_0^-)$ (resp. current of the inductor $i_L(t_0^-)$) previous to the transient beginning (instant t_0^-). This value is, in a first approximation, the only one remaining unchanged from t_0^- to t_0^+. With this value, the circuit can be analyzed at $t = 0^+$, where the capacitor (resp. inductor) is replaced by a voltage source (resp. current source) of value $u_C(t_0^-)$ (resp. $i_L(t_0^-)$). Additionally, the independent electrical sources are replaced by dc sources with a constant value equal to that in $t = t_0^+$.

It should be remarked that the previous three steps are independent of each other, so that their order can be freely altered. With this methodology, all the terms involved in the complete response of the variable under study are determined.

Example

In the circuit represented in Fig. 1.15, the switch is closed at $t = 0$. Calculate the voltage $u_C(t)$ and the current $i(t)$ for $t > 0$, knowing that $u_C(0^-) = 4\,\text{V}$.

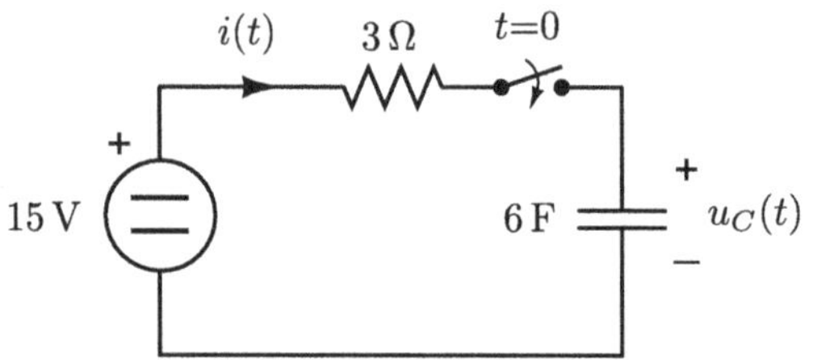

Fig. 1.15 Example circuit

Solution Being a first order circuit, the complete response of the voltage $u_C(t)$ and the current $i(t)$ can be expressed as:

$$u_C(t) = u_C^p(t) + \left[u_C(0^+) - u_C^p(0^+)\right] \cdot \mathrm{e}^{-t/\tau}$$
$$i(t) = i^p(t) + \left[i(0^+) - i^p(0^+)\right] \cdot \mathrm{e}^{-t/\tau}$$

The steady-state response, the time constant and the initial condition of these variables will be subsequently determined.

Steady-state response. In this case, after the switch is closed, only a dc source of excitation is present, the steady-state dc circuit shown in Fig. 1.16 is solved, where the capacitor has been replaced by an open circuit.

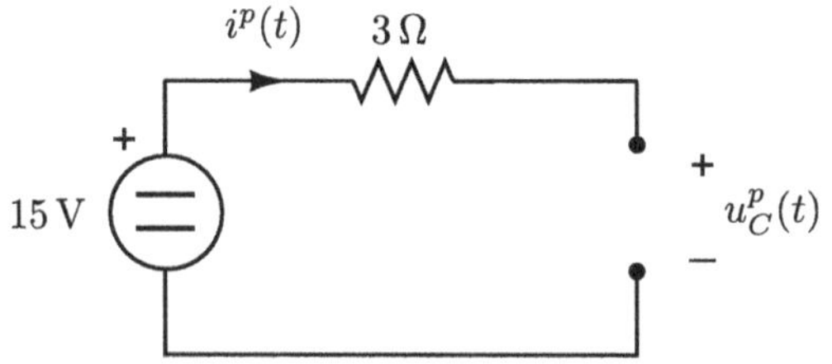

Fig. 1.16 Steady-state dc circuit

The resolution of the circuit in Fig. 1.16 leads to the following steady-state magnitudes:

$$i^p(t) = 0\,\text{A} \;\; ; \;\; u_C^p(t) = 15\,\text{V}$$

For $t = 0^+$:

$$i^p(0^+) = 0\,\text{A} \;\; ; \;\; u_C^p(0^+) = 15\,\text{V}$$

Time constant. Being an RC type circuit, the time constant is calculated as follows:

$$\tau = RC = 3 \cdot 6 = 18\,\text{s}$$

Initial conditions. According to the example statement, the capacitor voltage before the switch closure is $u_C(0^-) = 4\,\text{V}$. To obtain the initial conditions, the circuit will be studied at the instant $t = 0^+$. This circuit is obtained from the original one, once the switch is closed, substituting the capacitor for a dc voltage source with a value of $u_C(0^+)$. Except for an impulse-type response, the capacitor voltage does not change when the switch is closed, so that:

$$u_C(0^+) = u_C(0^-) = 4\,\text{V}$$

The circuit at time $t = 0^+$ is depicted in Fig. 1.17, from which it is easily obtained $i(0^+)$:

$$i(0^+) = \frac{15 - 4}{3} = \frac{11}{3}\,\text{A}$$

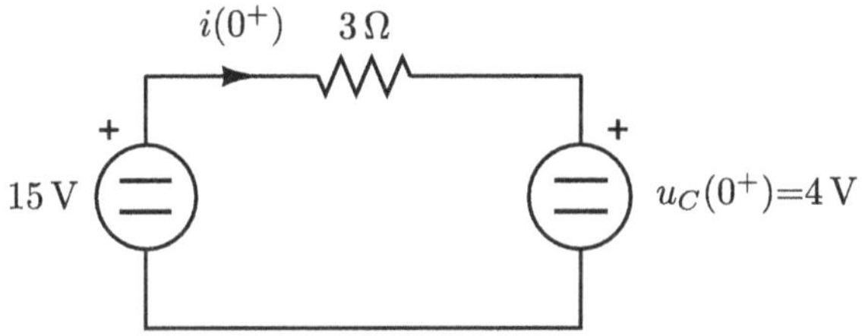

Fig. 1.17 Circuit at $t = 0^+$

The values of $i(t)$ and $u_C(t)$ for $t > 0$ are consequently calculated, yielding:

$$\begin{aligned} u_C(t) &= u_C^p(t) + \left[u_C(0^+) - u_C^p(0^+)\right] \cdot \mathrm{e}^{-t/\tau} \\ &= 15 + [4 - 15] \cdot \mathrm{e}^{-t/18} = 15 - 11 \cdot \mathrm{e}^{-t/18}\,\text{V} \end{aligned}$$

$$i(t) = i^p(t) + \left[i(0^+) - i^p(0^+)\right] \cdot \mathrm{e}^{-t/\tau} = 0 + \left[\frac{11}{3} - 0\right] \cdot \mathrm{e}^{-t/18} = \frac{11}{3} \cdot \mathrm{e}^{-t/18}\,\text{A}$$

Example

In the circuit represented in Fig. 1.18, the switch is closed at $t = 0$. Calculate the current $i_R(t)$ for $t > 0$, knowing that $i_g(t) = 20\sqrt{2}\sin(5t + 45°)\,\text{A}$.

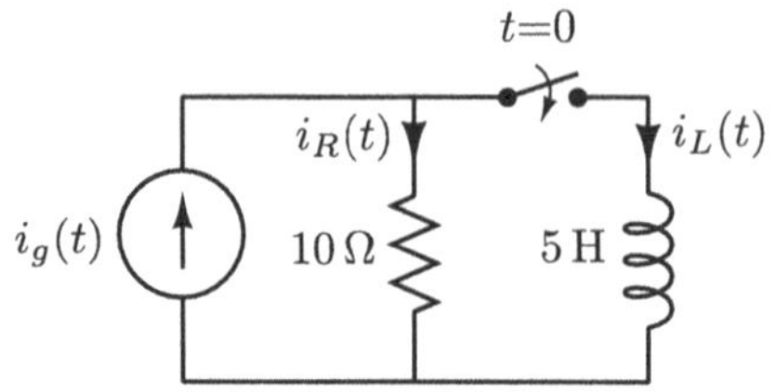

Fig. 1.18 Example circuit

Solution Being a first order circuit, the evolution of the current $i_R(t)$ is given by:

$$i_R(t) = i_R^p(t) + \left[i_R(0^+) - i_R^p(0^+)\right] \cdot \mathrm{e}^{-t/\tau}$$

The steady-state response, the time constant and the initial condition of this variable will be subsequently determined.

Steady-state response. In this case, once the switch is closed, only an ac source of excitation is present in the resulting circuit. For this reason, the steady-state ac regime will be solved, using the circuit depicted in Fig. 1.19 in the phasor domain.

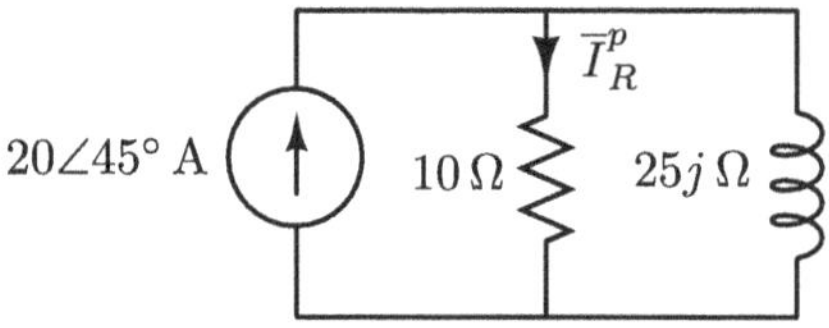

Fig. 1.19 Circuit in the phasor domain

The resolution of the circuit in Fig. 1.19 leads to:

$$\overline{I}_R^p = 20\angle 45^\circ \frac{\frac{1}{10}}{\frac{1}{10} + \frac{1}{25j}} = 20\angle 45^\circ \frac{25j}{10 + 25j} \approx 18{,}57\angle 66{,}8^\circ \text{ A}$$

Once the steady-state current is obtained in the phasor domain, the corresponding expression in the time domain is derived:

$$i_R^p(t) = 18{,}57\sqrt{2}\sin(5t + 66{,}8^\circ) \text{ A}$$

whose value for $t = 0^+$ yields:

$$i_R^p(0^+) = 18{,}57\sqrt{2}\sin(66{,}8^\circ) \approx 24{,}14 \text{ A}$$

Time constant. Being an RL circuit, the time constant (for $t > 0$) is given by:

$$\tau = \frac{L}{R} = \frac{5}{10} = \frac{1}{2} \text{ s}$$

Initial conditions. Before the switch closure, no current flows through the inductor, so that $i_L(0^-) = 0\,\text{A}$. To obtain the initial condition, the circuit corresponding to the instant $t = 0^+$ will be considered, being this circuit obtained from the original one, after the switch is closed, replacing the inductor with a dc current source of value $i_L(0^+)$. Except for an impulse-type response, the current of a inductor does not change when the switch is closed, therefore:

$$i_L(0^+) = i_L(0^-) = 0\,\text{A}$$

Additionally, the ac supply, $i_g(t)$, is replaced by a dc source with a constant value equal to $i_g(0^+)$, yielding:

$$i_g(0^+) = 20\sqrt{2}\sin(0 + 45^\circ) = 20\,\text{A}$$

The circuit at time $t = 0^+$ is shown in Fig. 1.20, from which the value of $i_R(0^+)$ is easily obtained:

$$i_R(0^+) = 20\,\text{A}$$

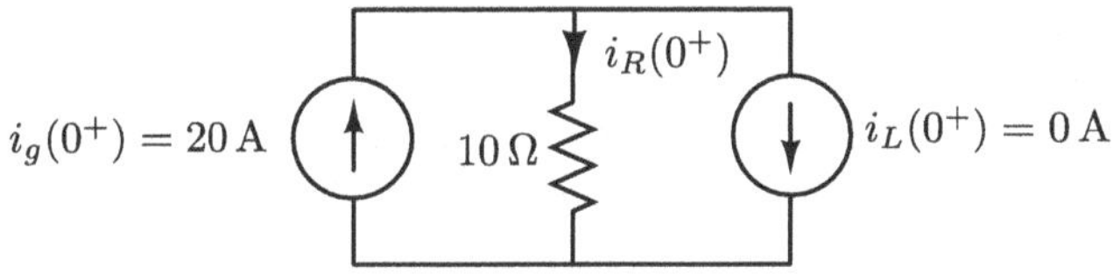

Fig. 1.20 Circuit at $t = 0^+$

Finally, the complete response of the current $i_R(t)$ for $t > 0$ is:

$$\begin{aligned} i_R(t) &= i_R^p(t) + \left[i_R(0^+) - i_R^p(0^+)\right] \cdot \text{e}^{-t/\tau} \\ &= 18{,}57\sqrt{2}\sin(5t + 66{,}8^\circ) + [20 - 24{,}14] \cdot \text{e}^{-2t}\,\text{A} \end{aligned}$$

1.6 Impulse Response

As previously mentioned, capacitor voltages and inductor currents cannot suffer discontinuities in absence of impulse-type responses. However, in certain situations these variables might suffer a jump discontinuity, meaning the appearance of a step in their time evolution, related to an impulse response. Those situations are mainly:

- Parallel connection of capacitors with different initial voltage.
- Series connection of inductors with different initial current.

1.6.1 Parallel Connection of Capacitors

Consider the circuit in Fig. 1.21, where the capacitors C_1 and C_2 will be connected in parallel when the switch closes at $t = t_0$. At instant t_0^-, just before the switch closing, the capacitors are charged at a different voltage, $u_1(t_0^-) \neq u_2(t_0^-)$. After the closure, both capacitors are in parallel and their voltages will necessarily be equal, that is, $u_1(t_0^+) = u_2(t_0^+) = u(t_0^+)$. In this particular case, it cannot be established that the voltage of each capacitor is unchanged from instant t_0^- to t_0^+, as it was assumed in the previous examples.

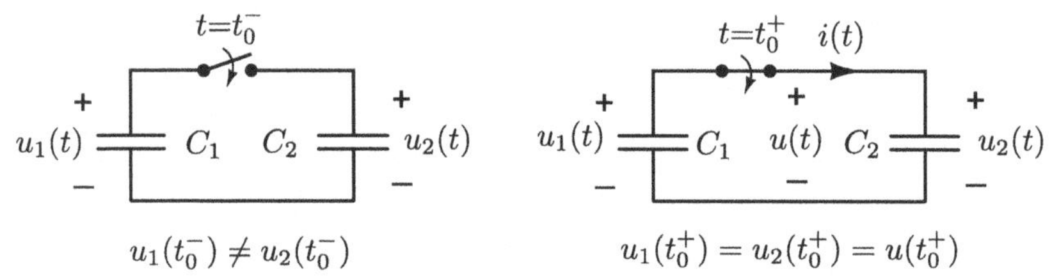

Fig. 1.21 Parallel connection of capacitors

In order to obtain the value of the common voltage of the two capacitors at t_0^+, the principle of charge conservation is applied, which states that the total charge in an isolated system is constant. This fact implies that the total stored charge cannot change abruptly, leading to the expression bellow:

$$\Sigma q_i(t_0^-) = \Sigma q_i(t_0^+) \quad \Rightarrow \quad C_1 u_1(t_0^-) + C_2 u_2(t_0^-) = (C_1 + C_2)\, u(t_0^+) \tag{1.34}$$

Therefore, the voltage that both capacitors will have at t_0^+ will be:

$$u(t_0^+) = \frac{C_1 u_1(t_0^-) + C_2 u_2(t_0^-)}{C_1 + C_2} \tag{1.35}$$

From that moment on, it can be considered that the circuit has a single capacitor with $C_{eq} = C_1 + C_2$ and a voltage $u(t_0^+)$.

Although the total charge of the system is conserved, there is a charge transfer from one capacitor to another in an infinitesimal time. This can only be achieved if the circulating current is very high (impulse-type response). The current flowing at the instant of switch closing can be calculated taking into account the following equation for the capacitor C_2:

$$i(t) = C_2 \frac{du_2(t)}{dt} \quad \Rightarrow \quad i(t)dt = C_2 du_2(t) \tag{1.36}$$

The integration of both members between t_0^+ and t_0^- results in:

$$\int_{t_0^-}^{t_0^+} i(t)dt = C_2 \int_{t_0^-}^{t_0^+} du_2(t) = C_2 u(t_0^+) - C_2 u_2(t_0^-) \neq 0 \tag{1.37}$$

Considering that the Dirac delta function (or unit impulse) satisfies the following properties:

$$\delta(t-t_0)=\begin{cases}\infty & \text{if } t=t_0\\ 0 & \text{if } t\neq t_0\end{cases} \qquad \int_{t_0^-}^{t_0^+}\delta(t-t_0)dt=1 \tag{1.38}$$

it can be concluded that the current will be:

$$i(t)=C_2\left(u(t_0^+)-u_2(t_0^-)\right)\cdot\delta(t-t_0) \tag{1.39}$$

For a general case including n capacitors with different initial voltages, connected in parallel at instant t_0, the common voltage at t_0^+ is given by:

$$u(t_0^+)=\frac{\sum_{i=1}^{n}C_iu_i(t_0^-)}{\sum_{i=1}^{n}C_i} \tag{1.40}$$

From the voltage variation suffered by each capacitor, the current flowing through it at t_0 is determined as follows:

$$i_i(t)=C_i\left(u(t_0^+)-u_i(t_0^-)\right)\cdot\delta(t-t_0) \tag{1.41}$$

In the equation above, passive references have been assumed, that is, the calculated current is that flowing from the positive to the negative terminal of the voltage.

1.6.2 Series Connection of Inductors

In the circuit represented in Fig. 1.22 there are two inductors through which currents $i_1(t)$ and $i_2(t)$ flow. After the switch opens, the inductors will be connected in series, therefore, they will have the same current. At instant t_0^- the currents have different values, $i_1(t_0^-)\neq i_2(t_0^-)$. When the switch is opened, the inductors will not be able to maintain their respective current values since it must be satisfied that $i_1(t_0^+)=i_2(t_0^+)=i(t_0^+)$. In this case, it cannot be considered that the inductor currents are unchanged from instant t_0^- to t_0^+.

To obtain the current value immediately after the switch opening, the principle of flux conservation will be considered, which establishes that the total flux cannot change abruptly:

$$\Sigma\phi_i(t_0^-)=\Sigma\phi_i(t_0^+)\ \Rightarrow\ L_1i_1(t_0^-)+L_2i_2(t_0^-)=(L_1+L_2)\,i(t_0^+) \tag{1.42}$$

Therefore, the current through both inductors at t_0^+ will be:

$$i(t_0^+)=\frac{L_1i_1(t_0^-)+L_2i_2(t_0^-)}{L_1+L_2} \tag{1.43}$$

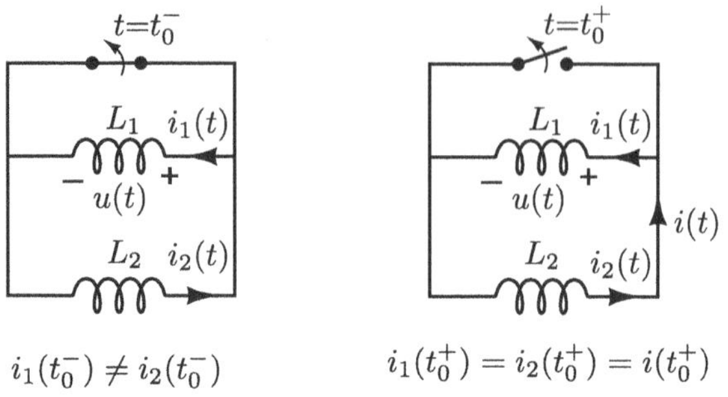

Fig. 1.22 Series connection of inductors

From that moment on, it can be considered that the circuit has a single inductor with a self-induction coefficient $L_{eq} = L_1 + L_2$ and an initial current $i(t_0^+)$.

The discontinuity appearing in the current of each inductor is associated with an impulse-type response in the corresponding voltage. This magnitude, for the inductor L_1, can be calculated using its defining equation:

$$u(t) = L_1 \frac{di_1(t)}{dt} \quad \Rightarrow \quad u(t)dt = L_1 di_1(t) \tag{1.44}$$

The integration of both members between t_0^+ and t_0^- results in:

$$\int_{t_0^-}^{t_0^+} u(t)dt = L_1 \int_{t_0^-}^{t_0^+} di_1(t) = L_1 i(t_0^+) - L_1 i_1(t_0^-) \neq 0 \tag{1.45}$$

Considering the properties of the Dirac delta function (1.38), the voltage value will be:

$$u(t) = L_1 \left(i(t_0^+) - i_1(t_0^-) \right) \cdot \delta(t - t_0) \tag{1.46}$$

This result can easily be extended to n inductors with different initial currents, connected in series at a given time t_0. The current through all of them at t_0^+ is given by:

$$i(t_0^+) = \frac{\sum_{i=1}^{n} L_i i_i(t_0^-)}{\sum_{i=1}^{n} L_i} \tag{1.47}$$

Finally, from the variation suffered by the current of each inductor, the corresponding voltage can be determined at t_0 as:

$$u_i(t) = L_i \left(i(t_0^+) - i_i(t_0^-) \right) \cdot \delta(t - t_0) \tag{1.48}$$

In the equation above, passive references have been assumed for the voltage and current of each inductor.

1.7 Solved Problems

1.7.1 RC Under DC Supply

In the circuit represented in Fig. 1.23, the capacitor is discharged when the switch is closed at $t = 0$. Calculate the voltage $u_C(t)$ and the current $i(t)$ for $t > 0$.

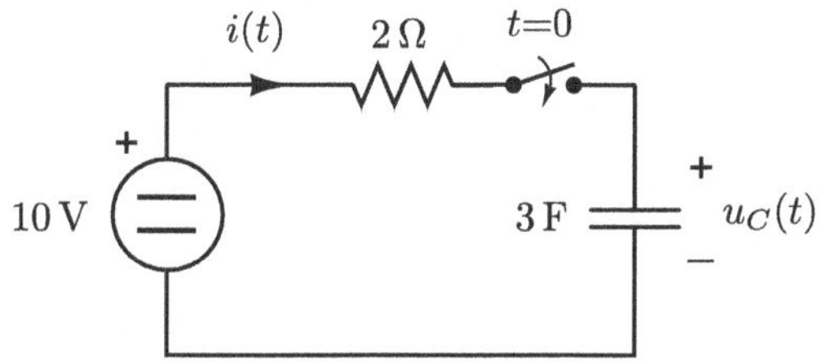

Fig. 1.23 Circuit considered in the problem under study

Solution. According to the problem statement, before the switch closing, that is, at $t = 0^-$, the capacitor is discharged, meaning that $u_C(0^-) = 0\,\text{V}$. To obtain the initial conditions, the circuit at $t = 0^+$ will be used, which is obtained from the original circuit, once the switch is closed, replacing the capacitor with a dc voltage source of $u_C(0^+)$. Except for an impulse-type response, the capacitor voltage does not change when the switch is closed, therefore:

$$u_C(0^+) = u_C(0^-) = 0\,\text{V}$$

The circuit at time $t = 0^+$ is shown in Fig. 1.24, from which $i(0^+)$ is easily obtained:

$$i(0^+) = \frac{10}{2} = 5\,\text{A}$$

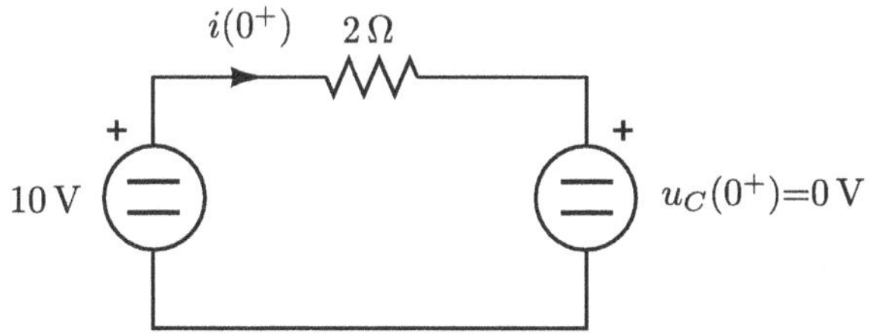

Fig. 1.24 Circuit at $t = 0^+$

A few important steady-state magnitudes will be subsequently calculated. After the switch is closed, only a dc-type source of excitation is present in the resulting circuit, so that the dc circuit shown in Fig. 1.25 is solved, where the capacitor has been replaced by an open circuit.

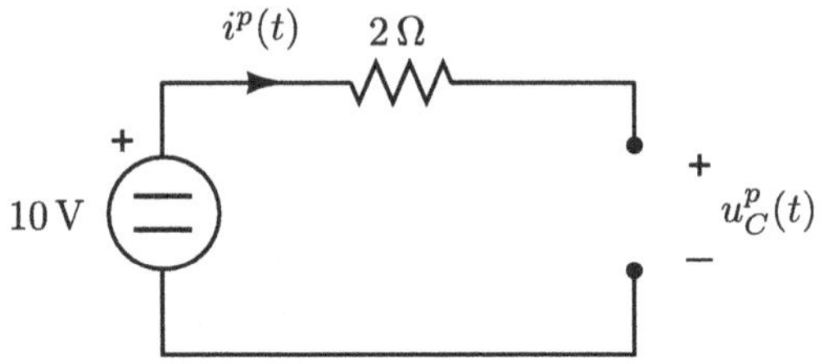

Fig. 1.25 Circuit in dc steady-state regime

The resolution of the circuit in Fig. 1.25 leads to the following steady-state values:

$$i^p(t) = 0\,\text{A} \ ; \ u_C^p(t) = 10\,\text{V}$$

Substituting $t = 0^+$:

$$i^p(0^+) = 0\,\text{A} \ ; \ u_C^p(0^+) = 10\,\text{V}$$

Being an RC type circuit, the time constant is obtained as follows:

$$\tau = RC = 2 \cdot 3 = 6\,\text{s}$$

Finally, the complete responses of $i(t)$ and $u_C(t)$ for $t > 0$ are consequently determined:

$$\begin{aligned} i(t) &= i^p(t) + \left[i(0^+) - i^p(0^+)\right] \cdot \text{e}^{-t/\tau} \\ &= 0 + [5 - 0] \cdot \text{e}^{-t/6} = 5 \cdot \text{e}^{-t/6}\,\text{A} \\ u_C(t) &= u_C^p(t) + \left[u_C(0^+) - u_C^p(0^+)\right] \cdot \text{e}^{-t/\tau} \\ &= 10 + [0 - 10] \cdot \text{e}^{-t/6} = 10 - 10 \cdot \text{e}^{-t/6}\,\text{V} \end{aligned}$$

1.7.2 RL Under DC Supply

In the circuit shown in Fig. 1.26, the switch is closed at $t = 0$. Calculate the current $i_L(t)$ and voltage $u_L(t)$ for $t > 0$.

Solution. Before the switch closing, no current flows through the inductor, so that $i_L(0^-) = 0\,\text{A}$. To obtain the initial conditions, the circuit at $t = 0^+$ will be considered, which is obtained from the original one, after the switch is closed. The inductor is replaced by a dc current source of value $i_L(0^+)$. Except for an impulse-type response, the inductor current does not change when the switch is closed, therefore:

$$i_L(0^+) = i_L(0^-) = 0\,\text{A}$$

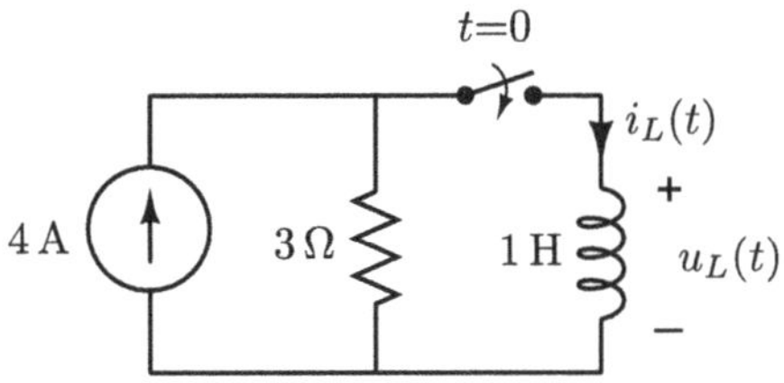

Fig. 1.26 Circuit considered in the problem under study

The circuit at $t = 0^+$ is shown in Fig. 1.27.

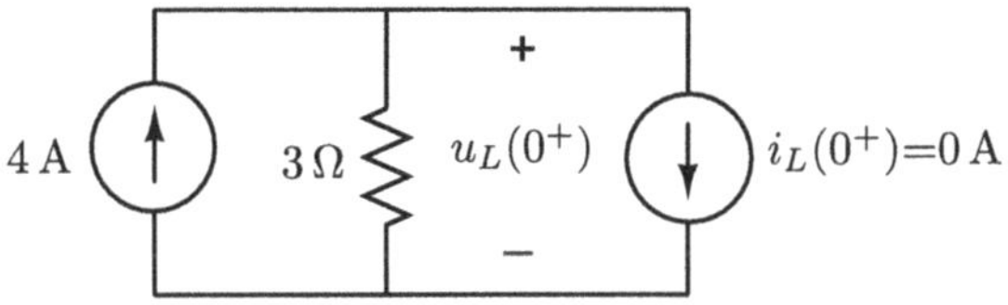

Fig. 1.27 Circuit at $t = 0^+$

From the circuit in Fig. 1.27 the value of $u_L(0^+)$ is easily obtained:

$$u_L(0^+) = 4 \cdot 3 = 12\,\text{V}$$

The steady-state values of the magnitudes under study will be subsequently obtained. In this case, since only a dc-type source of excitation is present, the dc circuit shown in Fig. 1.28 will be solved, where the inductor has been replaced by a short circuit.

Solving the circuit in Fig. 1.28, the following steady-state values can be obtained:

$$i_L^p(t) = 4\,\text{A} \;\; ; \;\; u_L^p(t) = 0\,\text{V}$$

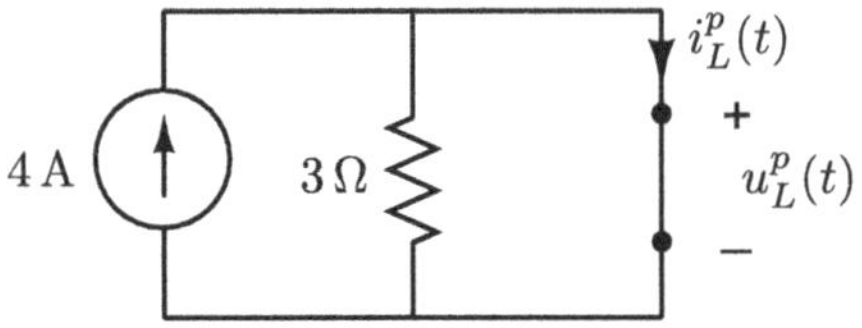

Fig. 1.28 Circuit in dc steady-state regime

Substituting $t = 0^+$:

$$i_L^p(0^+) = 4\,\text{A} \;\; ; \;\; u_L^p(0^+) = 0\,\text{V}$$

Being an RL type circuit, the time constant is calculated as follows:

$$\tau = \frac{L}{R} = \frac{1}{3}\,\mathrm{s}$$

Finally, the complete responses of $i_L(t)$ and $u_L(t)$ for $t > 0$ are consequently determined:

$$i_L(t) = i_L^p(t) + \left[i_L(0^+) - i_L^p(0^+)\right] \cdot \mathrm{e}^{-t/\tau} = 4 + [0 - 4] \cdot \mathrm{e}^{-3t} = 4 \cdot (1 - \mathrm{e}^{-3t})\,\mathrm{A}$$

$$u_L(t) = u_L^p(t) + \left[u_L(0^+) - u_L^p(0^+)\right] \cdot \mathrm{e}^{-t/\tau} = 0 + [12 - 0] \cdot \mathrm{e}^{-3t} = 12 \cdot \mathrm{e}^{-3t}\,\mathrm{V}$$

The time evolution of $i_L(t)$ and $u_L(t)$ is graphically represented in Fig. 1.29.

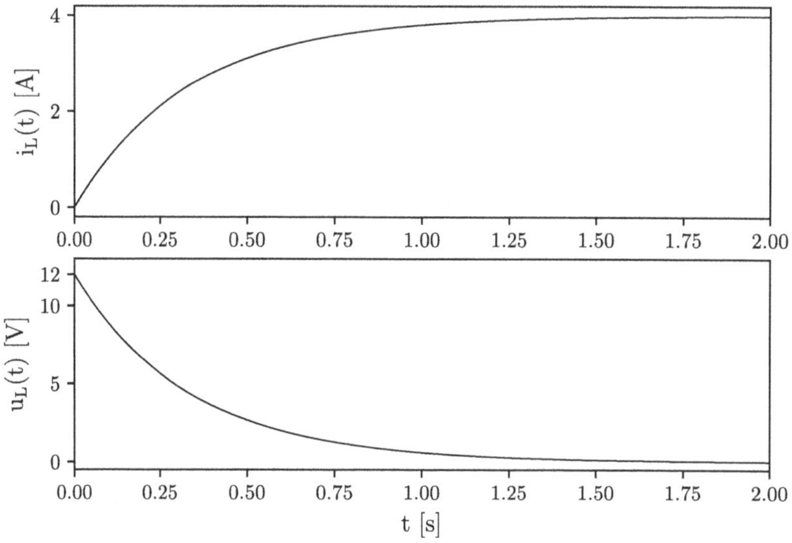

Fig. 1.29 Time evolution of $i_L(t)$ and $u_L(t)$

1.7.3 RC Without Excitation Sources

The circuit represented in Fig. 1.30 is in steady state when the switch is opened at $t = 0$. Calculate the current $i_C(t)$ for $t > 0$.

Solution. First, the capacitor voltage will be obtained before the switch opening, that is, at time $t = 0^-$. For this purpose, it will be considered that the circuit is in steady-state regime before the switch is opened. Given that only a dc-type source of excitation is present, the dc circuit shown in Fig. 1.31 will be solved, where the capacitor has been replaced by an open circuit.

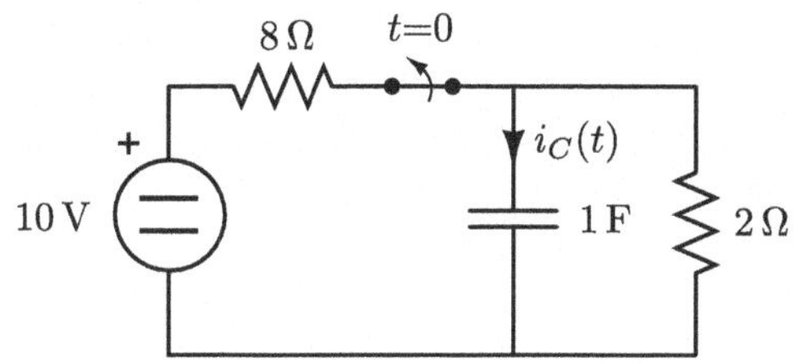

Fig. 1.30 Circuit considered in the problem under study

From Fig. 1.31, the value of the capacitor voltage at $t = 0^-$ can be obtained as follows:

$$u_C(0^-) = 10 \cdot \frac{2}{8+2} = 2\,\text{V}$$

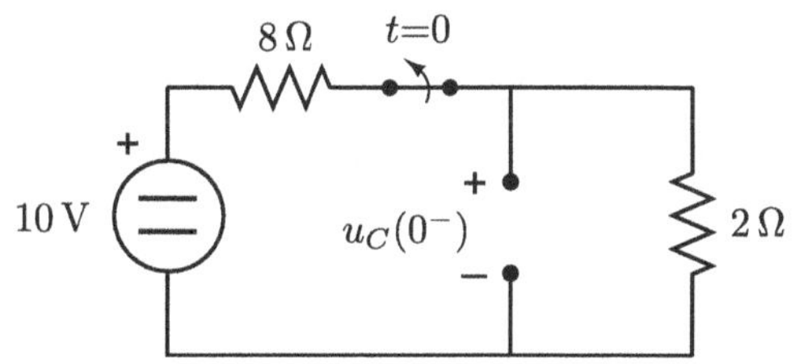

Fig. 1.31 Circuit at $t = 0^-$

Except for an impulse-type response, the capacitor voltage does not change when the switch is opened, therefore:

$$u_C(0^+) = u_C(0^-) = 2\,\text{V}$$

To obtain the initial condition of the current $i(t)$, the circuit at instant $t = 0^+$ will be considered, which is obtained from the original one, once the switch is opened, replacing the capacitor with a dc voltage source of value $u_C(0^+)$. The circuit at time $t = 0^+$ is shown in Fig. 1.32.

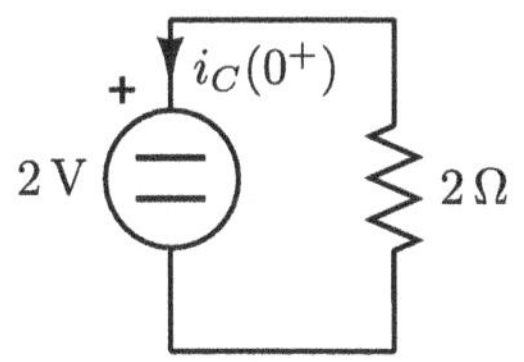

Fig. 1.32 Circuit at $t = 0^+$

From the circuit in Fig. 1.32, the value of $i_C(0^+)$ can be calculated as:

$$i_C(0^+) = \frac{-2}{2} = -1\,\text{A}$$

The steady-state value of the current $i_C(t)$ is subsequently obtained. In this case, after the switch is open, there is no source of excitation in the resulting circuit, remaining exclusively the capacitor with the resistor connected. As the steady state corresponds to the response forced by the sources of excitation, if there is no source of excitation, there is no steady-state regime. Thus:

$$i_C^p(t) = 0\,\mathrm{A} \;\; ; \;\; i_C^p(0^+) = 0\,\mathrm{A}$$

Being an RC circuit, the value of the time constant (for $t > 0$) is calculated as follows:

$$\tau = RC = 2 \cdot 1 = 2\,\mathrm{s}$$

Finally, after the initial condition, the steady-state response and the time constant have been obtained, the complete response of $i_C(t)$ is derived below for $t > 0$:

$$i_C(t) = i_C^p(t) + \left[i_C(0^+) - i_C^p(0^+)\right] \cdot \mathrm{e}^{-t/\tau} = 0 + [-1 - 0] \cdot \mathrm{e}^{-t/2} = -1 \cdot \mathrm{e}^{-t/2}\,\mathrm{A}$$

The time evolution of $i_C(t)$ has been graphically represented in Fig. 1.33.

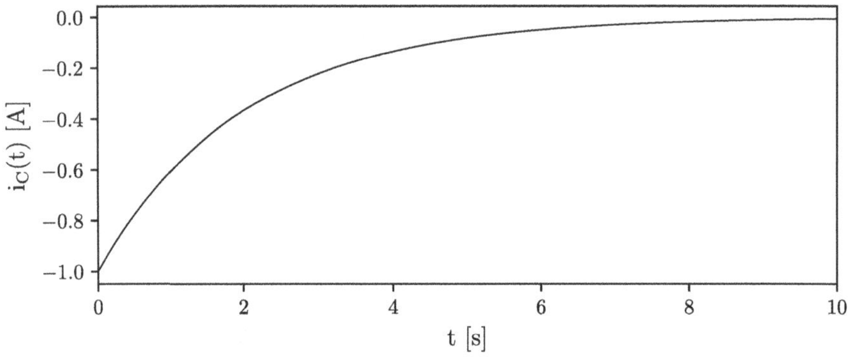

Fig. 1.33 Time evolution of $i_C(t)$

1.7.4 RL Without Excitation Sources

The circuit in Fig. 1.34 is in steady state when the switch is opened at $t = 0$. Calculate the current $i_L(t)$ and the voltage $u_L(t)$ for $t > 0$.

Solution. First, the current through the 5 H inductor will be obtained before the switch opening, that is, at $t = 0^-$. For this purpose, it will be considered that the circuit is in steady state before the switch is opened. Given that only a dc-type source of excitation is present, the dc circuit shown in Fig. 1.35 will be solved, where each inductor has been replaced by a short circuit.

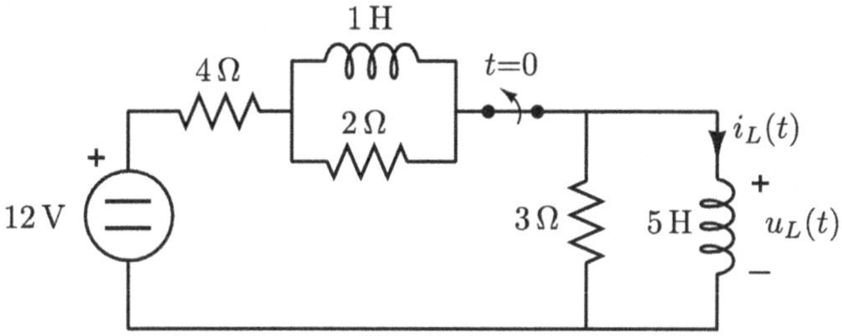

Fig. 1.34 Circuit considered in the problem under study

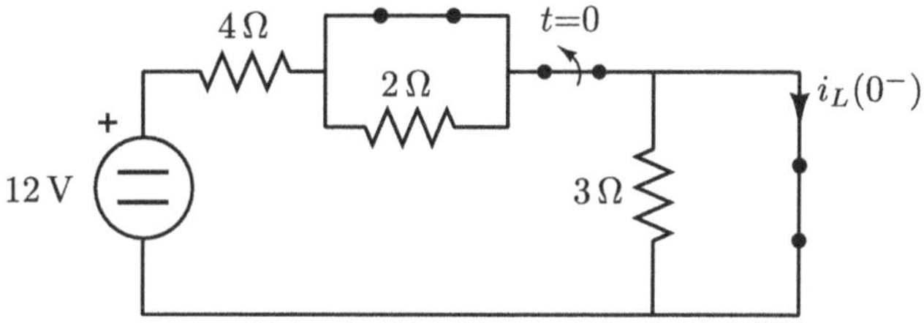

Fig. 1.35 Circuit at $t = 0^-$

From the circuit in Fig. 1.35, the inductor current is obtained at $t = 0^-$:

$$i_L(0^-) = \frac{12}{4} = 3\,\mathrm{A}$$

Except for an impulse-type response, the current through the inductor does not change when the switch is opened, therefore:

$$i_L(0^+) = i_L(0^-) = 3\,\mathrm{A}$$

To obtain the initial conditions of the current $i_L(t)$ and the voltage $u_L(t)$, the circuit at instant $t = 0^+$ will be used, which is obtained from the original one, once the switch has been opened, replacing the 5 H inductor with a dc current source of value $i_L(0^+)$. The resulting circuit at time $t = 0^+$ is shown in Fig. 1.36.

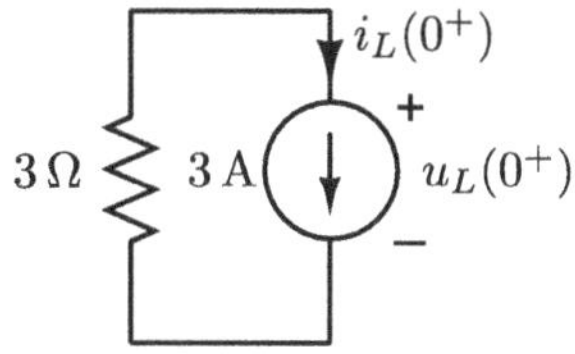

Fig. 1.36 Circuit at $t = 0^+$

From the circuit in Fig. 1.36, $i_L(0^+)$ and $u_L(0^+)$ are calculated as follows:

$$i_L(0^+) = 3\,\text{A} \ ; \quad u_L(0^+) = -3 \cdot 3 = -9\,\text{V}$$

The steady-state values of the current $i_L(t)$ and the voltage $u_L(t)$ are subsequently obtained. In this case, once the switch is open, there is no source of excitation in the resulting circuit. Consequently:

$$i_L^p(t) = 0\,\text{A} \ ; \quad u_L^p(t) = 0\,\text{V}$$
$$i_L^p(0^+) = 0\,\text{A} \ ; \quad u_L^p(0^+) = 0\,\text{V}$$

Being an RL circuit, the time constant (for $t > 0$) is obtained as:

$$\tau = \frac{L}{R} = \frac{5}{3}\,\text{s}$$

Finally, the complete response of $i_L(t)$ and $u_L(t)$ for $t > 0$ yields:

$$i_L(t) = i_L^p(t) + \left[i_L(0^+) - i_L^p(0^+)\right] \cdot \text{e}^{-t/\tau} = 0 + [3 - 0] \cdot \text{e}^{-3t/5} = 3 \cdot \text{e}^{-3t/5}\,\text{A}$$
$$u_L(t) = u_L^p(t) + \left[u_L(0^+) - u_L^p(0^+)\right] \cdot \text{e}^{-t/\tau} = 0 + [-9 - 0] \cdot \text{e}^{-3t/5} = -9 \cdot \text{e}^{-3t/5}\,\text{V}$$

1.7.5 RL Under DC Supply

The circuit represented in Fig. 1.37 is in steady state when the switch is opened at $t = 0$. Calculate the current $i(t)$ for $t > 0$.

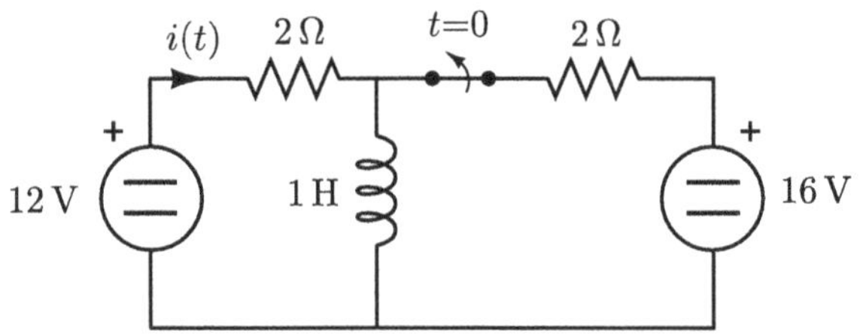

Fig. 1.37 Circuit considered in the problem under study

Solution. First, the inductor current will be obtained before the switch opening, that is, at instant $t = 0^-$. For this purpose, it will be considered that the circuit is in steady-state regime before the switch is opened. Given that both excitation sources are of dc voltage, the circuit shown in Fig. 1.38 will be solved, where the inductor has been replaced by a short circuit.

The inductor current at $t = 0^-$ is obtained from Fig. 1.38:

$$i_L(0^-) = \frac{12}{2} + \frac{16}{2} = 6 + 8 = 14\,\text{A}$$

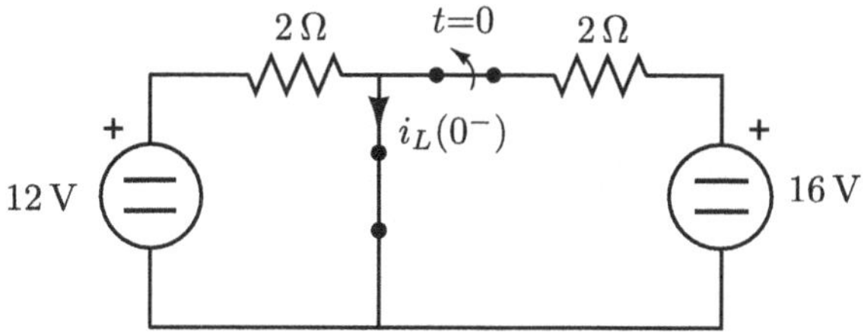

Fig. 1.38 Circuit at $t = 0^-$

Except for an impulse-type response, the inductor current does not change when the switch is opened, therefore:

$$i_L(0^+) = i_L(0^-) = 14\,\text{A}$$

To calculate the initial value of the current $i(t)$, the circuit at instant $t = 0^+$ will be used, which is obtained from the original one, once the switch has been opened, replacing the inductor with a dc current source of value $i_L(0^+)$. The circuit at $t = 0^+$ is shown in Fig. 1.39.

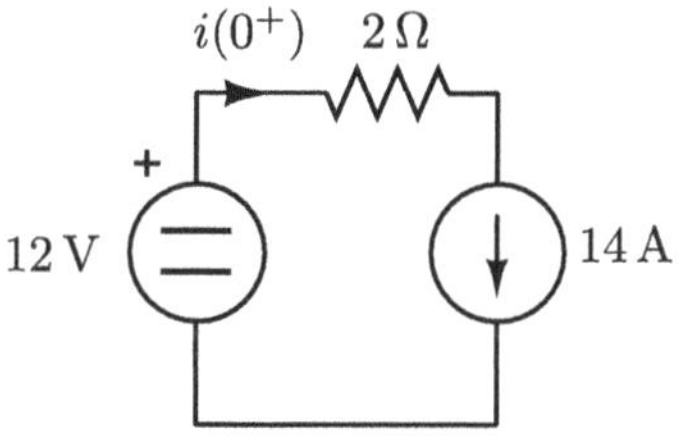

Fig. 1.39 Circuit at $t = 0^+$

From the circuit in Fig. 1.39, the current $i(0^+)$ can be obtained:

$$i(0^+) = 14\,\text{A}$$

The steady-state value of the current $i(t)$ is subsequently obtained. In this case, once the switch is open, only a dc-type source of excitation is present in the resulting circuit. For this reason, the circuit shown in Fig. 1.40 will be solved, where the inductor has been replaced by a short circuit.

From the circuit in Fig. 1.40:

$$i^p(t) = \frac{12}{2} = 6\,\text{A}$$

Substituting $t = 0^+$:

$$i^p(0^+) = 6\,\text{A}$$

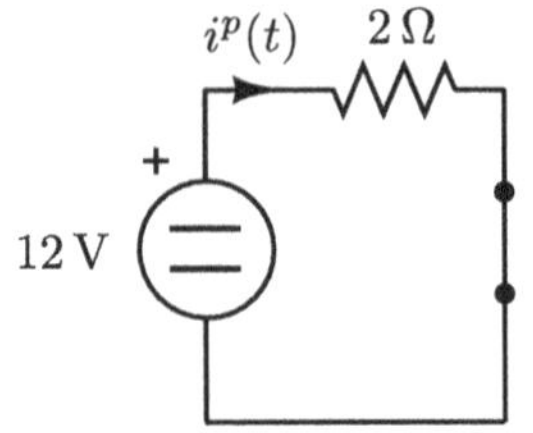

Fig. 1.40 Circuit in dc steady-state regime

Being an RL circuit, the time constant (for $t > 0$) is calculated as follows:

$$\tau = \frac{L}{R} = \frac{1}{2}\ \text{s}$$

Finally, once the initial condition, the steady-state response and the time constant have been calculated, the complete response of $i(t)$ for $t > 0$ is obtained as:

$$i(t) = i^p(t) + \left[i(0^+) - i^p(0^+)\right] \cdot \mathrm{e}^{-t/\tau} = 6 + [14 - 6] \cdot \mathrm{e}^{-2t} = 6 + 8 \cdot \mathrm{e}^{-2t}\ \text{A}$$

1.7.6 RC Under DC Supply

The circuit represented in Fig. 1.41 is in steady state when, at $t = 0$, the switch k1 is closed and k2 is opened. Calculate the current $i(t)$ for $t > 0$.

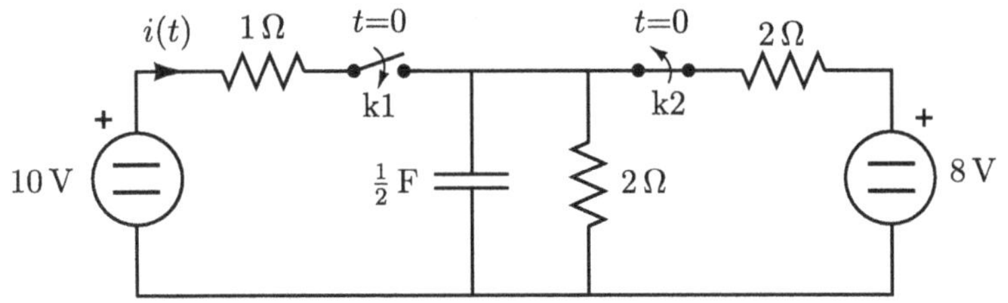

Fig. 1.41 Circuit considered in the problem under study

Solution. First, the capacitor voltage will be obtained before the switches change their position, that is, at instant $t = 0^-$. For this purpose, it will be considered that the circuit is in steady state before switch k1 closes and switch k2 opens. Given that, in this situation, only a dc-type source of excitation is present, the dc circuit shown in Fig. 1.42 will be solved, where the capacitor has been replaced by an open circuit.

From the circuit in Fig. 1.42, the value of the capacitor voltage at $t = 0^-$ can be obtained as:

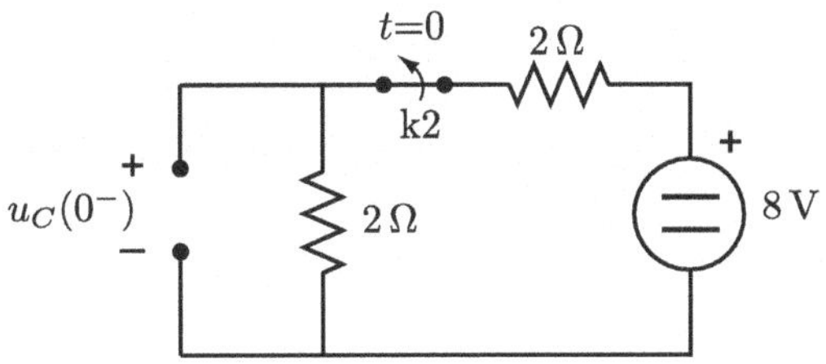

Fig. 1.42 Circuit at $t = 0^-$

$$u_C(0^-) = 8 \cdot \frac{2}{2+2} = 4\,\text{V}$$

Except for an impulse-type response, which is not the case, the capacitor voltage does not change when switch k1 closes and switch k2 opens, therefore:

$$u_C(0^+) = u_C(0^-) = 4\,\text{V}$$

To obtain the initial value of the current $i(t)$, the circuit at instant $t = 0^+$ will be used, which is obtained from the original one, once the switches change their position, replacing the capacitor with a voltage source of value $u_C(0^+)$. The circuit at time $t = 0^+$ is shown in Fig. 1.43.

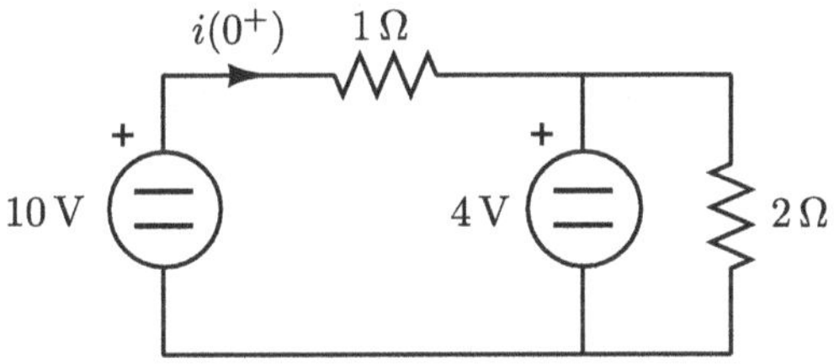

Fig. 1.43 Circuit at $t = 0^+$

From the circuit in Fig. 1.43, the current $i(0^+)$ can be obtained:

$$i(0^+) = \frac{10 - 4}{1} = 6\,\text{A}$$

The steady-state value of the current $i(t)$ is subsequently obtained. In this case, once switch k1 closes and switch k2 opens, only a dc-type source of excitation is present in the resulting circuit. For this reason, the dc circuit shown in Fig. 1.44 will be solved.

From the circuit in Fig. 1.44:

$$i^p(t) = \frac{10}{1+2} = \frac{10}{3}\,\text{A}$$

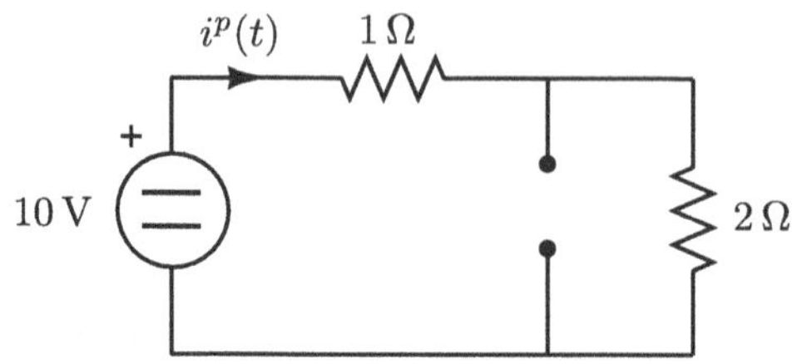

Fig. 1.44 Circuit in dc steady-state regime

Substituting $t = 0^+$:

$$i^p(0^+) = \frac{10}{3} \text{ A}$$

Being an RC circuit, the time constant is calculated as:

$$\tau = R_{eq}C$$

where R_{eq} is the equivalent resistance from the capacitor terminals, corresponding to the parallel connection of the 1 Ω resistance and the 2 Ω resistance, yielding:

$$\tau = R_{eq}C = \frac{2 \cdot 1}{2+1} \cdot \frac{1}{2} = \frac{1}{3} \text{ s}$$

Finally, once the initial condition, the steady-state response and the time constant have been obtained, the complete response of $i(t)$ for $t > 0$ is derived below:

$$i(t) = i^p(t) + \left[i(0^+) - i^p(0^+)\right] \cdot \mathrm{e}^{-t/\tau} = \frac{10}{3} + \left[6 - \frac{10}{3}\right] \cdot \mathrm{e}^{-3t} = \frac{10}{3} + \frac{8}{3} \cdot \mathrm{e}^{-3t} \text{ A}$$

The time evolution of $i(t)$ is graphically represented in Fig. 1.45.

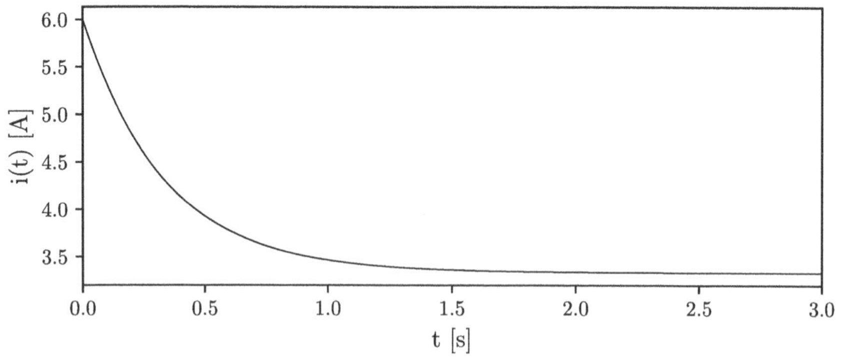

Fig. 1.45 Time evolution of $i(t)$

1.7.7 RL Under DC Supply

The circuit represented in Fig. 1.46 is in steady state when, at $t = 0$, the switch k1 is closed and the k2 is opened. Calculate the current $i_L(t)$ for $t > 0$, knowing that $e(t) = 100\sqrt{2}\sin(50t)$ V.

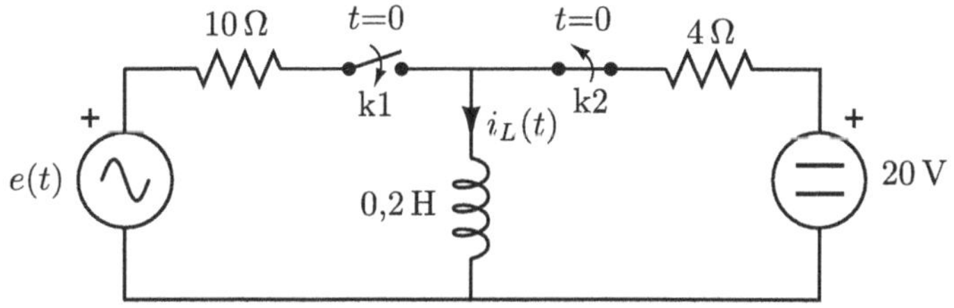

Fig. 1.46 Circuit considered in the problem under study

Solution. First, the inductor current will be obtained before the switches change their position ($t = 0^-$). For this purpose, it will be considered that the circuit is in steady state before this time. Given that, in this situation, only a dc-type source of excitation is present, the dc circuit shown in Fig. 1.47 will be solved, where the inductor has been replaced by a short circuit.

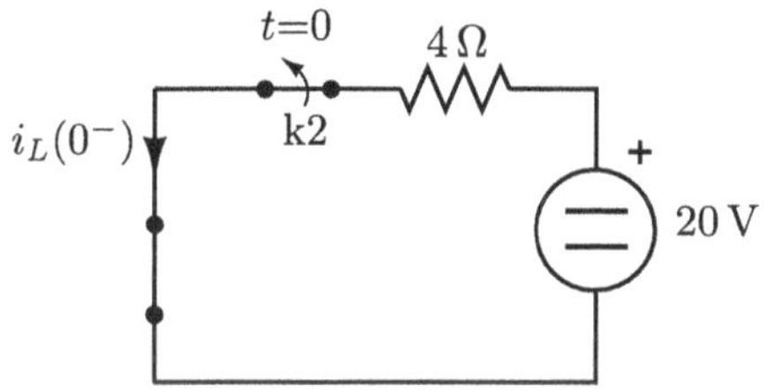

Fig. 1.47 Circuit at $t = 0^-$

From the circuit in Fig. 1.47, the inductor current at $t = 0^-$ is obtained as:

$$i_L(0^-) = \frac{20}{4} = 5\,\text{A}$$

Except for an impulse-type response, which is not the case, the inductor current does not change when switch k1 closes and switch k2 opens, therefore:

$$i_L(0^+) = i_L(0^-) = 5\,\text{A}$$

In this particular case, it is not necessary to solve the circuit at $t = 0^+$, since the only significant magnitude is the inductor current, whose value has already been determined.

The steady-state value of the current $i_L(t)$ is subsequently obtained. In this case, once switch k1 closes and switch k2 opens, only an ac-type source of excitation is present in the resulting circuit. For this reason, the circuit in the phasor domain shown in Fig. 1.48 will be considered.

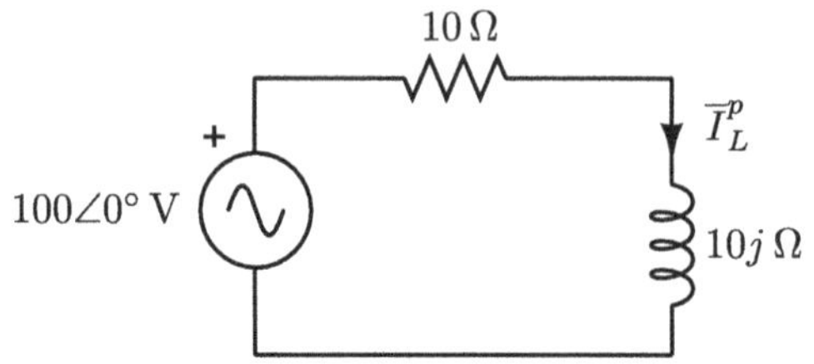

Fig. 1.48 Circuit in phasor domain

From the circuit in Fig. 1.48:

$$\overline{I}_L^p = \frac{100\angle 0°}{10 + 10j} = \frac{10}{\sqrt{2}}\angle - 45° \text{ A}$$

After the steady-state current has been obtained in the phasor domain, the corresponding time domain expression is derived:

$$i_L^p(t) = 10 \sin(50t - 45°) \text{ A}$$

whose value for $t = 0^+$ is:

$$i_L^p(0^+) = 10 \sin(-45°) = \frac{-10 \cdot \sqrt{2}}{2} = -5\sqrt{2} \text{ A}$$

Being an RL circuit, the time constant (for $t > 0$) is calculated as:

$$\tau = \frac{L}{R} = \frac{0{,}2}{10} = \frac{1}{50} \text{ s}$$

Finally, once the initial condition, the steady-state response and the time constant have been obtained, the complete response of $i_L(t)$ is obtained below for $t > 0$:

$$i_L(t) = i_L^p(t) + \left[i_L(0^+) - i_L^p(0^+)\right] \cdot \mathrm{e}^{-t/\tau} = 10 \sin(50t - 45°) + \left[5 - (-5\sqrt{2})\right] \cdot \mathrm{e}^{-50t} \text{ A}$$

The time evolution of $i_L(t)$ is graphically represented in Fig. 1.49.

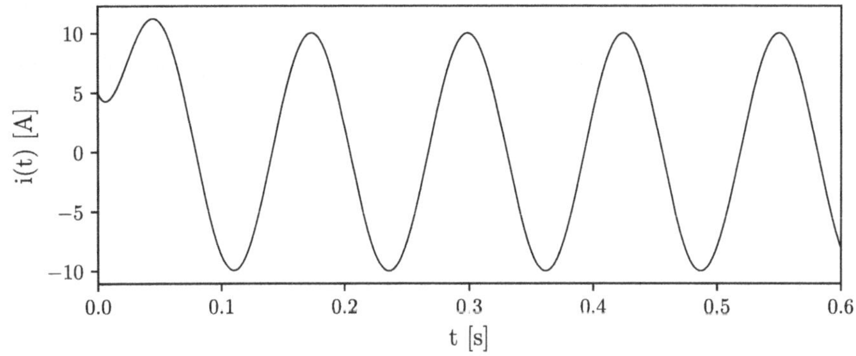

Fig. 1.49 Time evolution of $i_L(t)$

1.7.8 RL Under DC Supply

In the circuit depicted in Fig. 1.50, the switch is closed at $t = 0$. Calculate the current $i_L(t)$ for $t > 0$, knowing that $e(t) = 150\sin(500t)$ V.

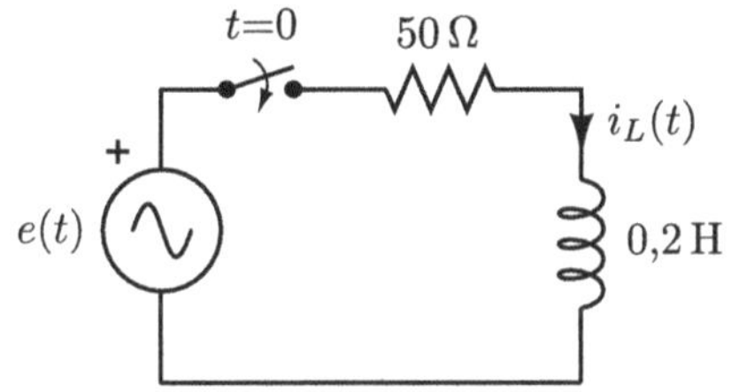

Fig. 1.50 Circuit considered in the problem under study

Solution. Before the switch closing, it is known that $i_L(0^-) = 0$ A. Except for an impulse-type response, the inductor current does not change when the switch closes, therefore:

$$i_L(0^+) = i_L(0^-) = 0\,\text{A}$$

The steady-state value of the inductor current, $i_L^p(t)$, is subsequently obtained. In this case, since only an ac-type source of excitation is present, the circuit in the phasor domain shown in Fig. 1.51 will be solved.

From the circuit in Fig. 1.51:

$$\overline{I}_L^p = \frac{(150/\sqrt{2})\angle 0^\circ}{50 + 100j} \approx 0{,}95\angle - 63{,}43^\circ\,\text{A}$$

Once the steady-state current is obtained in the phasor domain, the corresponding expression in the time domain is derived:

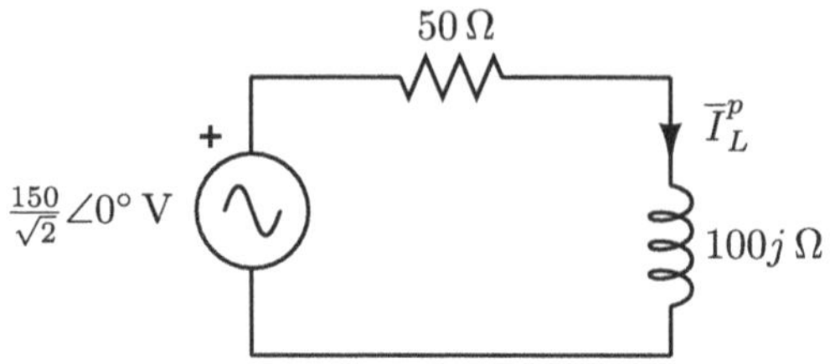

Fig. 1.51 Circuit in the phasor domain

$$i_L^p(t) = 0{,}95\sqrt{2}\sin(500t - 63{,}43°) \approx 1{,}34\sin(500t - 63{,}43°)\text{ A}$$

whose value for $t = 0^+$ is:

$$i_L^p(0^+) = 1{,}34\sin(-63{,}43°) \approx -1{,}2\text{ A}$$

Being an RL type circuit, the time constant is calculated as follows:

$$\tau = \frac{L}{R} = \frac{0{,}2}{50} = \frac{1}{250}\text{ s}$$

Finally, the complete response of $i_L(t)$ for $t > 0$ is consequently obtained:

$$i_L(t) = i_L^p(t) + \left[i_L(0^+) - i_L^p(0^+)\right] \cdot \mathrm{e}^{-t/\tau} = 1{,}34\sin(500t - 63{,}43°) + [0 + 1{,}2] \cdot \mathrm{e}^{-250t}\text{ A}$$

The time evolution of $i_L(t)$ is graphically represented in Fig. 1.52.

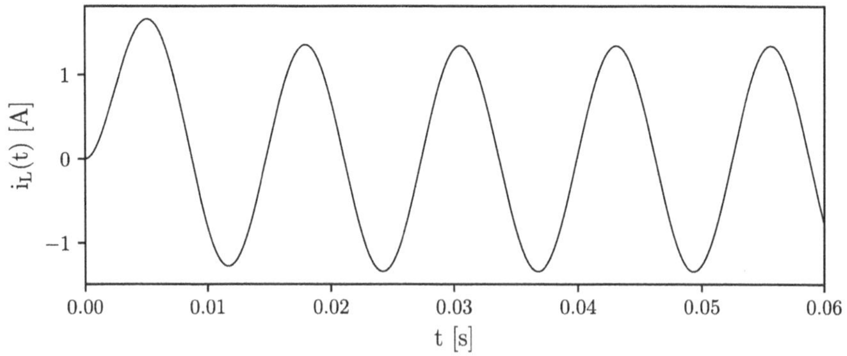

Fig. 1.52 Time evolution of $i_L(t)$

1.7.9 RC Under DC Supply

The circuit represented in Fig. 1.53 is in steady state when, at $t = 0$, the switch k1 is closed and the k2 is opened. Calculate the current $i(t)$ for $t > 0$, knowing that $e(t) = 6\cos(10t)$ V.

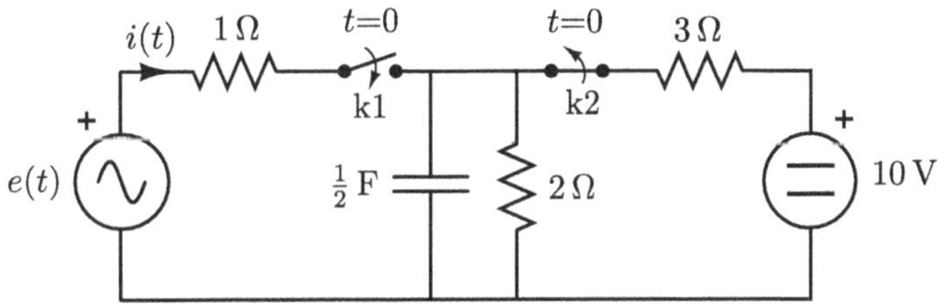

Fig. 1.53 Circuit considered in the problem under study

Solution. First, the capacitor voltage will be obtained before the switches change their position, that is, at instant $t = 0^-$. For this purpose, it will be considered that the circuit is in steady state before switch k1 closes and switch k2 opens. Given that, in this situation, only a dc-type source of excitation is present, the circuit shown in Fig. 1.54 will be solved, where the capacitor has been replaced by an open circuit.

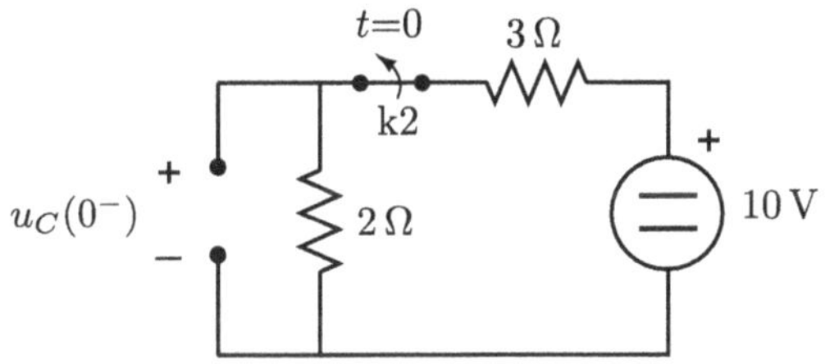

Fig. 1.54 Circuit at $t = 0^-$

From the circuit in Fig. 1.54, the capacitor voltage at $t = 0^-$ can be obtained:

$$u_C(0^-) = 10 \cdot \frac{2}{2+3} = 4\,\text{V}$$

Except for an impulse-type response, which is not the case, the capacitor voltage does not change when switch k1 closes and switch k2 opens, therefore:

$$u_C(0^+) = u_C(0^-) = 4\,\text{V}$$

To obtain the initial value of the current $i(t)$, the circuit at instant $t = 0^+$ will be used, which is obtained from the original one, once the switches change their position, replacing the capacitor with a dc voltage source of value $u_C(0^+)$. The circuit at time $t = 0^+$ is shown in Fig. 1.55.

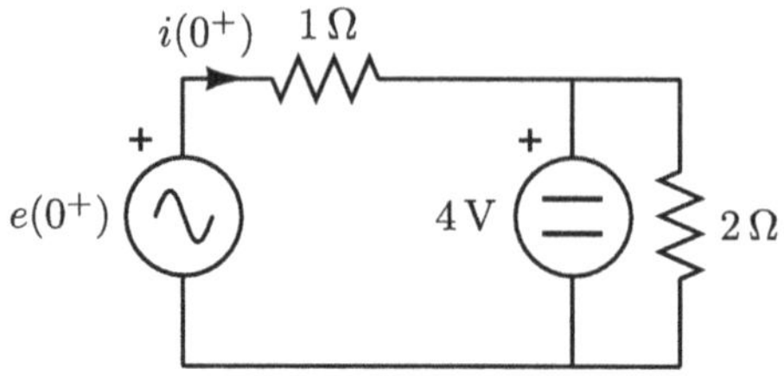

Fig. 1.55 Circuit at $t = 0^+$

From the circuit in Fig. 1.55, the value of $i(0^+)$ can be obtained as:

$$i(0^+) = \frac{e(0^+) - 4}{1} = \frac{6\cos(10 \cdot 0) - 4}{1} = 2\,\text{A}$$

The steady-state value of the current $i(t)$ is subsequently obtained. In this case, once switch k1 closes and switch k2 opens, only an ac-type source of excitation is present in the resulting circuit. For this reason, the circuit in the phasor domain shown in Fig. 1.56 will be solved.

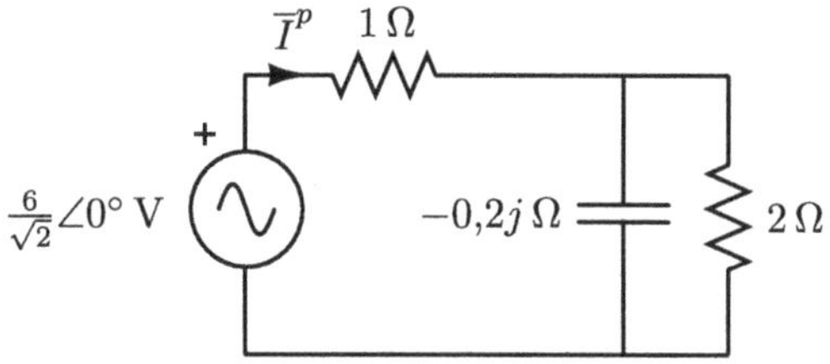

Fig. 1.56 Circuit in phasor domain

From the circuit in Fig. 1.56:

$$\overline{I}^p = \frac{(6/\sqrt{2})\angle 0^\circ}{1 + \frac{-0{,}2j \cdot 2}{2 - 0{,}2j}} \approx 4\angle 11^\circ\,\text{A}$$

Once the steady-state current is obtained in the phasor domain, the corresponding expression in the time domain is derived:

$$i^p(t) = 4\sqrt{2}\cos(10t + 11^\circ)\,\text{A}$$

whose value for $t = 0^+$ yields:

$$i^p(0^+) = 4\sqrt{2}\cos(11^\circ) \approx 5{,}55\,\text{A}$$

Being an RC circuit, the time constant (for $t > 0$) is calculated as:

$$\tau = R_{eq} C$$

where R_{eq} is the equivalent resistance from the capacitor terminals, corresponding to the parallel connection of the 1 Ω resistance and the 2 Ω resistance, yielding:

$$\tau = R_{eq} C = \frac{2 \cdot 1}{2+1} \cdot \frac{1}{2} = \frac{1}{3}\ \mathrm{s}$$

Finally, once the initial condition, the steady-state response and the time constant have been obtained, the complete response of $i(t)$ is obtained for $t > 0$:

$$i(t) = i^p(t) + \left[i(0^+) - i^p(0^+)\right] \cdot \mathrm{e}^{-t/\tau} = 4\sqrt{2}\cos(10t + 11^\circ) + [2 - 5{,}55] \cdot \mathrm{e}^{-3t}\ \mathrm{A}$$

The time evolution of $i(t)$ is graphically represented in Fig. 1.57.

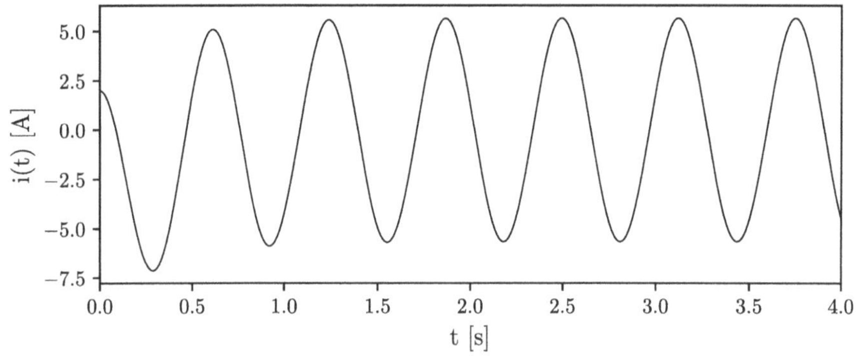

Fig. 1.57 Time evolution of $i(t)$

1.7.10 RC Under DC Supply

The switch in the circuit represented in Fig. 1.58 closes at $t = 0$. Given that the capacitor is discharged before this closing, calculate the voltage $u_C(t)$ for $t > 0$.

Solution. According to the problem statement, before the switch closing ($t < 0^-$), the capacitor is discharged, meaning that $u_C(0^-) = 0\,\mathrm{V}$. Except for an impulse-type response, the capacitor voltage does not change when the switch is closed, therefore:

$$u_C(0^+) = u_C(0^-) = 0\,\mathrm{V}$$

The steady-state value of the voltage $u_C(t)$ is subsequently obtained. In this case, once the switch is closed, only a dc-type source of excitation is present in the resulting

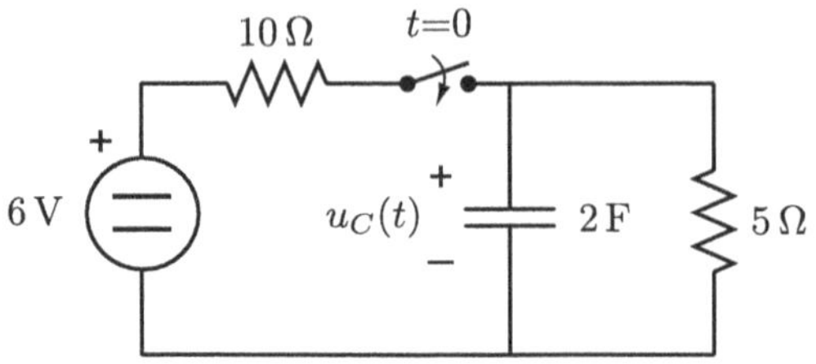

Fig. 1.58 Circuit considered in the problem under study

circuit. For this reason, the dc circuit shown in Fig. 1.59 will be solved, where the capacitor has been replaced by an open circuit.

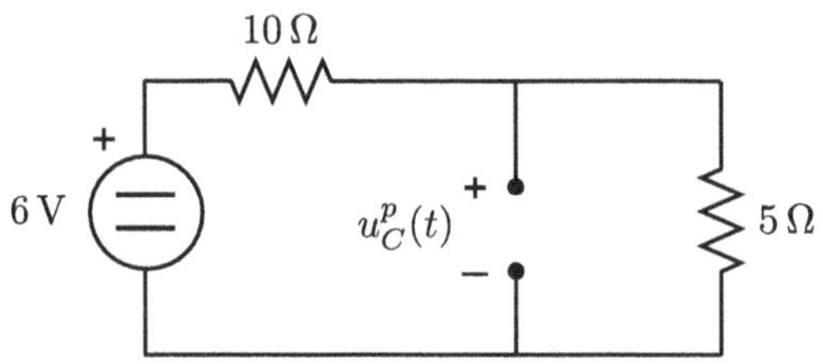

Fig. 1.59 Circuit in dc steady-state regime

From the circuit in Fig. 1.59:

$$u_C^p(t) = 6 \cdot \frac{5}{10+5} = 2\,\mathrm{V}$$

Substituting $t = 0^+$:

$$u_C^p(0^+) = 2\,\mathrm{V}$$

Being an RC circuit, the time constant (for $t > 0$) is calculated as follows:

$$\tau = R_{eq} C$$

where R_{eq} is the equivalent resistance from the capacitor terminals, corresponding to the parallel connection of the 10 Ω resistance and the 5 Ω resistance, yielding:

$$\tau = R_{eq} C = \frac{10 \cdot 5}{10+5} \cdot 2 = \frac{20}{3}\,\mathrm{s}$$

Finally, once the initial condition, the steady-state response and the time constant have been obtained, the complete response of $u_C(t)$ is obtained below for $t > 0$:

$$u_C(t) = u_C^p(t) + \left[u_C(0^+) - u_C^p(0^+)\right] \cdot \mathrm{e}^{-t/\tau} = 2 + [0-2] \cdot \mathrm{e}^{-3t/20} = 2 - 2 \cdot \mathrm{e}^{-3t/20}\,\mathrm{V}$$

The time evolution of $u_C(t)$ is graphically represented in Fig. 1.60.

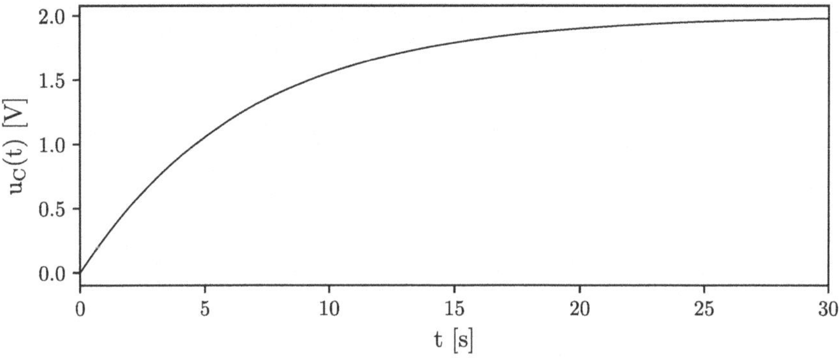

Fig. 1.60 Time evolution of $u_C(t)$

1.7.11 RL Under DC Supply

The switch in the circuit represented in Fig. 1.61 is closed at $t = 0$. Calculate the voltage $u_L(t)$ for $t > 0$, knowing that $i_L(0^-) = 0$ A.

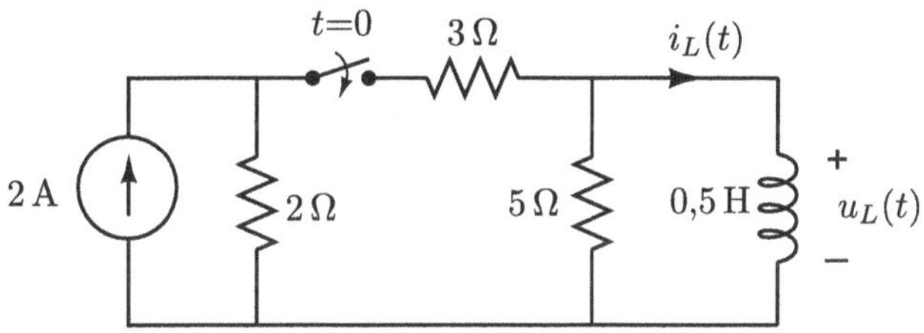

Fig. 1.61 Circuit considered in the problem under study

Solution. According to the problem statement, before the switch closing the value of the current is $i_L(0^-)$=0 A. To obtain the initial condition, the circuit at instant $t = 0^+$ will be used, which is obtained from the original one, once the switch is closed, replacing the inductor with a dc current source of value $i_L(0^+)$. Except for an impulse-type response, the inductor current does not change when the switch is closed, therefore:

$$i_L(0^+) = i_L(0^-) = 0\,\text{A}$$

The circuit at time $t = 0^+$ is shown in Fig. 1.62.

From the circuit depicted in Fig. 1.62 it can be noticed that $i_L(0^+) = 0$, so that:

$$i_1(0^+) = i_2(0^+)$$

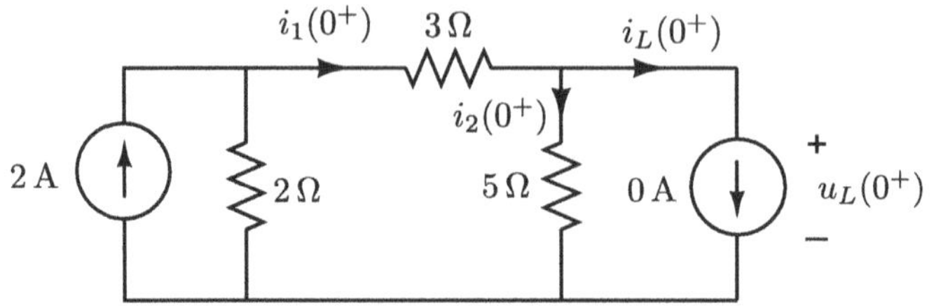

Fig. 1.62 Circuit at $t = 0^+$

Using the well-known concept of current divider, the following result is obtained:

$$i_2(0^+) = 2 \cdot \frac{\frac{1}{8}}{\frac{1}{8} + \frac{1}{2}} = 0{,}4\,\text{A}$$

Thus:

$$u_L(0^+) = 5 \cdot i_2(0^+) = 5 \cdot 0{,}4 = 2\,\text{V}$$

The steady-state value of the voltage $u_L(t)$ is subsequently obtained. In this case, once the switch is closed, only a dc-type source of excitation is present in the resulting circuit. Given that an inductor behaves like a short circuit in steady-state dc regime, the following is derived:

$$u_L^p(t) = 0\,\text{V} \ ; \ u_L^p(0^+) = 0\,\text{V}$$

Being an RL circuit, the time constant (for $t > 0$) is calculated as:

$$\tau = \frac{L}{R_{eq}}$$

where R_{eq} is the equivalent resistance from the inductor terminals (a and b). The corresponding passive circuit is shown in Fig. 1.63.

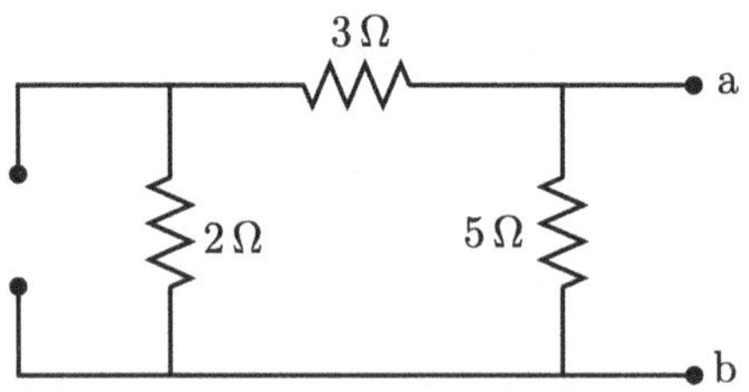

Fig. 1.63 Passive circuit from inductor terminals

From the circuit represented in Fig. 1.63:

$$R_{eq} = 5//(2+3) = \frac{5 \cdot (2+3)}{5 + (2+3)} = \frac{5}{2}\,\Omega$$

Yielding:

$$\tau = \frac{L}{R_{eq}} = \frac{0{,}5}{5/2} = \frac{1}{5}\,\mathrm{s}$$

Finally, once the initial condition, the steady-state response and the time constant have been obtained, the complete response of the voltage $u_L(t)$ is obtained below for $t > 0$:

$$u_L(t) = u_L^p(t) + \left[u_L(0^+) - u_L^p(0^+)\right] \cdot \mathrm{e}^{-t/\tau} = 0 + [2 - 0] \cdot \mathrm{e}^{-5t} = 2 \cdot \mathrm{e}^{-5t}\,\mathrm{V}$$

The time evolution of $u_L(t)$ is graphically represented in Fig. 1.64.

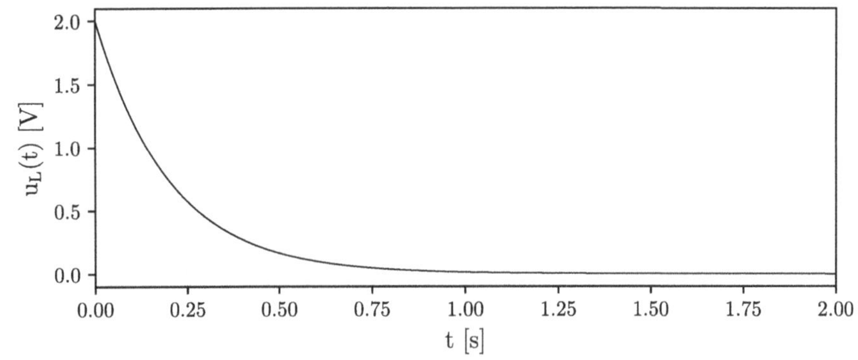

Fig. 1.64 Time evolution of $u_L(t)$

1.7.12 RC Under DC and AC Supply

In the circuit represented in Fig. 1.65, the switches are closed at $t = 0$. Given that $u_C(0^-) = 5\,\mathrm{V}$ and $e(t) = 6\cos(10t)\,\mathrm{V}$, calculate the current $i(t)$ and the voltage $u_C(t)$ for $t > 0$.

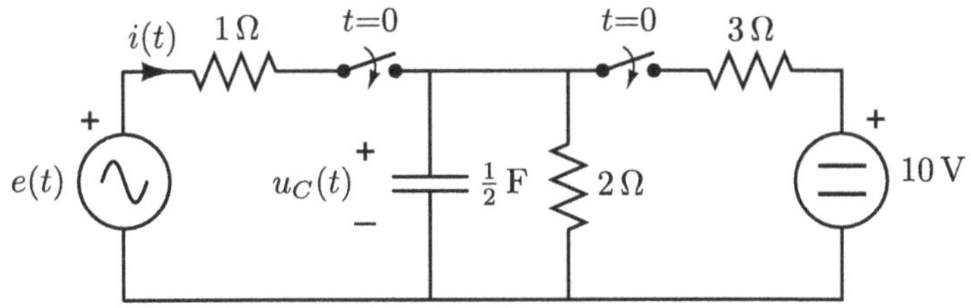

Fig. 1.65 Circuit considered in the problem under study

Solution. Except for an impulse-type response, which is not the case, the capacitor voltage does not change when the switches are closed, therefore:

$$u_C(0^+) = u_C(0^-) = 5\,\text{V}$$

To obtain the initial value of the current $i(t)$ the circuit at the instant $t = 0^+$ will be used, which is obtained from the original one, once both switches have been closed, substituting the capacitor for a voltage source with a value of $u_C(0^+)$, and the electrical sources for their values at $t = 0^+$. The circuit at instant $t = 0^+$ is shown in Fig. 1.66, from which the current $i(0^+)$ can be obtained as:

$$i(0^+) = \frac{e(0^+) - 5}{1} = \frac{6\cos(10 \cdot 0) - 5}{1} = 1\,\text{A}$$

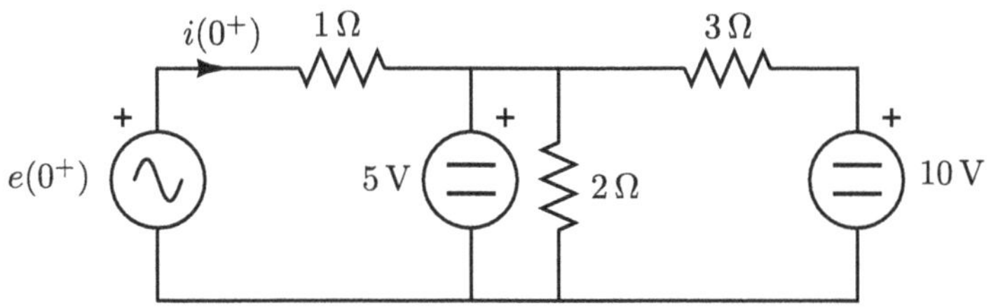

Fig. 1.66 Circuit at $t = 0^+$

The steady-state values of the current $i(t)$ and the voltage $u_C(t)$ are subsequently obtained. In this case, once the switches are closed, there are an ac voltage source and a dc voltage source. For this reason and in order to obtain the circuit resolution, the superposition principle must be applied. This way, the resulting circuit when only the dc voltage source is active is that shown in Fig. 1.67, where the capacitor has been replaced by an open circuit and the ac voltage source has been replaced by a short circuit.

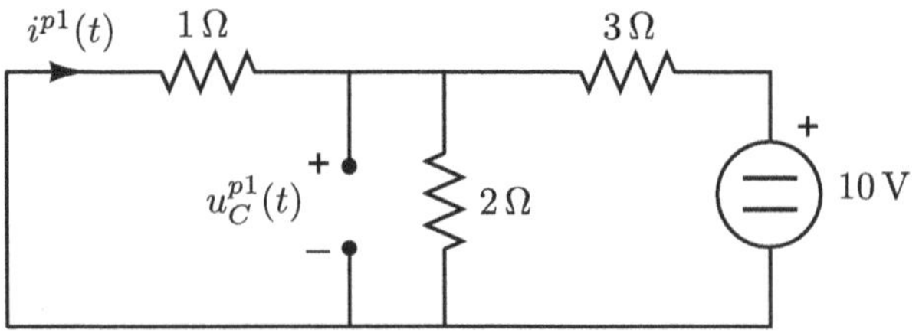

Fig. 1.67 Circuit in dc steady-state regime

From the circuit represented in Fig. 1.67:

$$i^{p1}(t) = -\frac{10}{3} \cdot \frac{1}{1+\frac{1}{3}+\frac{1}{2}} = \frac{-20}{11}\ \text{A}$$
$$u_C^{p1}(t) = -i^{p1} \cdot 1 = \frac{20}{11} \cdot 1 = \frac{20}{11}\ \text{V}$$

On the other hand, when only the ac source is active, the circuit will be solved in steady-state ac regime. To approach this resolution, the circuit in the phasor domain shown in Fig. 1.68 will be used:

$$\overline{Z}_{eq1} = 2//3 = \frac{2 \cdot 3}{2+3} = \frac{6}{5}\ \Omega$$

$$\overline{Z}_{eq2} = \overline{Z}_{eq1}//(-0{,}2j) = \frac{\overline{Z}_{eq1} \cdot (-0{,}2j)}{\overline{Z}_{eq1} + (-0{,}2j)} \approx 0{,}197\angle - 80{,}53°\ \Omega$$

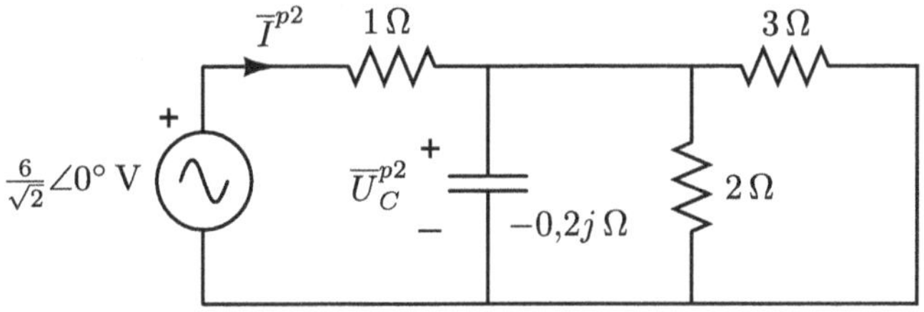

Fig. 1.68 Circuit in phasor domain

Therefore:

$$\overline{I}^{p2} = \frac{(6/\sqrt{2})\angle 0°}{1+\overline{Z}_{eq2}} \approx 4{,}04\angle 10{,}67°\ \text{A}$$
$$\overline{U}_C^{p2} = (6/\sqrt{2})\angle 0° - 1 \cdot \overline{I}^{p2} \approx 0{,}8\angle - 69{,}86°\ \text{V}$$

Once the steady-state magnitudes have been obtained in the phasor domain, the corresponding expressions in the time domain will be obtained:

$$i^{p2}(t) = 4{,}04\sqrt{2}\cos(10t + 10{,}67°)\ \text{A}$$
$$u_C^{p2}(t) = 0{,}8\sqrt{2}\cos(10t - 69{,}86°)\ \text{V}$$

Using the previous results:

$$i^{p}(t) = i^{p1}(t) + i^{p2}(t) = \frac{-20}{11} + 4{,}04\sqrt{2}\cos(10t + 10{,}67°)\ \text{A}$$
$$u_C^{p}(t) = u_C^{p1}(t) + u_C^{p2}(t) = \frac{20}{11} + 0{,}8\sqrt{2}\cos(10t - 69{,}86°)\ \text{V}$$

substituting $t = 0^+$:

$$i^p(0^+) = \frac{-20}{11} + 4{,}04\sqrt{2}\cos(10{,}67°) \approx 3{,}8\,\text{A}$$
$$u_C^p(0^+) = \frac{20}{11} + 0{,}8\sqrt{2}\cos(-69{,}86°) \approx 2{,}21\,\text{V}$$

Being an RC circuit, the time constant (for $t > 0$) is calculated as:

$$\tau = R_{eq}C$$

where R_{eq} is the equivalent resistance from the capacitor terminals, corresponding to the parallel connection of the 1 Ω resistance, that of 2 Ω and that of 3 Ω, yielding:

$$\tau = R_{eq}C = \frac{1}{\frac{1}{1} + \frac{1}{2} + \frac{1}{3}} \cdot \frac{1}{2} = \frac{3}{11}\,\text{s}$$

Finally, the complete response of $i(t)$ and $u_C(t)$ for $t > 0$ are consequently derived:

$$\begin{aligned} i(t) &= i^p(t) + \left[i(0^+) - i^p(0^+)\right] \cdot \mathrm{e}^{-t/\tau} \\ &= \frac{-20}{11} + 4{,}04\sqrt{2}\cos(10t + 10{,}67°) + [1 - 3{,}8] \cdot \mathrm{e}^{-11t/3}\,\text{A} \\ u_C(t) &= u_C^p(t) + \left[u_C(0^+) - u_C^p(0^+)\right] \cdot \mathrm{e}^{-t/\tau} \\ &= \frac{20}{11} + 0{,}8\sqrt{2}\cos(10t - 69{,}86°) + [5 - 2{,}21] \cdot \mathrm{e}^{-11t/3}\,\text{V} \end{aligned}$$

The time evolution of $i(t)$ and $u_C(t)$ is graphically represented in Fig. 1.69.

1.7.13 RC. Capacitors in Parallel Under DC Supply

In the circuit represented in Fig. 1.70, the switch closes at $t = 0$. Knowing that $u(0^-) = 6\,\text{V}$, calculate the voltage $u(t)$ for $t > 0$.

Solution. From the circuit in Fig. 1.70, it can be observed the parallel connection of two capacitors with an initial charging voltage equal to 6 V. This configuration is equivalent to a single capacitor with a charging voltage of 6 V, whose equivalent capacity is given by:

$$C_{eq} = C_1 + C_2 = 7 + 3 = 10\,\mu\text{F}$$

This way, the circuit is shown in Fig. 1.71.

Except for an impulse-type response, the capacitor voltage does not change when the switch is closed, therefore:

$$u(0^+) = u(0^-) = 6\,\text{V}$$

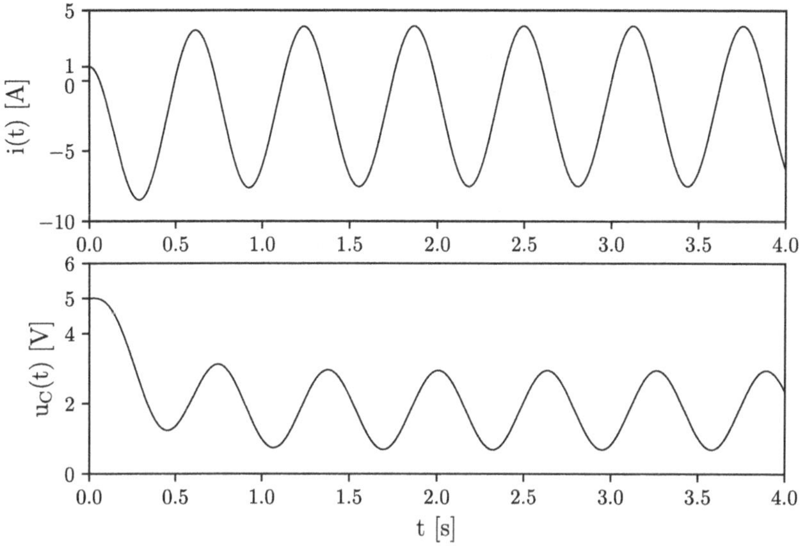

Fig. 1.69 Time evolution of $i(t)$ y $u_C(t)$

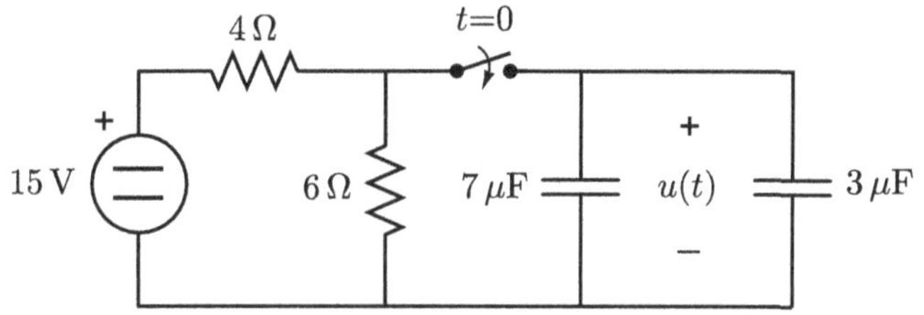

Fig. 1.70 Circuit considered in the problem under study

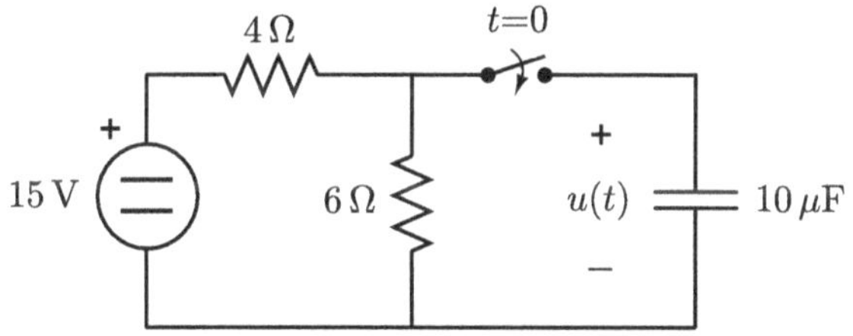

Fig. 1.71 Simplified circuit

The steady-state value of the voltage $u(t)$ is subsequently obtained. In this case, once the switch is closed, only a dc-type source of excitation is present in the resulting circuit. For this reason, the dc circuit shown in Fig. 1.72 will be solved, where the capacitor has been replaced by an open circuit.

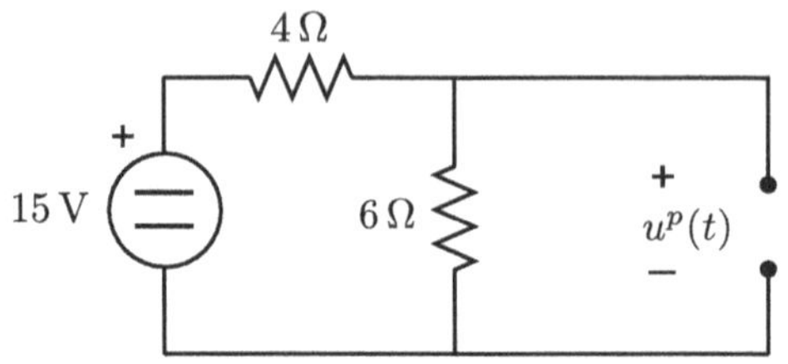

Fig. 1.72 Circuit in dc steady-state regime

From the circuit represented in Fig. 1.72:

$$u^p(t) = 15 \cdot \frac{6}{4+6} = 9\,\text{V}$$

Substituting $t = 0^+$:

$$u^p(0^+) = 9\,\text{V}$$

Being an RC circuit, the time constant is calculated as follows:

$$\tau = R_{eq} C_{eq}$$

where R_{eq} is the equivalent resistance from the capacitor terminals, corresponding to the parallel connection of the 4 Ω resistance and the 6 Ω resistance, yielding:

$$\tau = R_{eq} C_{eq} = \frac{4 \cdot 6}{4+6} \cdot 10 \cdot 10^{-6} = 24 \cdot 10^{-6}\,\text{s}$$

Finally, the complete response of $u(t)$ for $t > 0$ is consequently derived:

$$u(t) = u^p(t) + \left[u(0^+) - u^p(0^+)\right] \cdot \mathrm{e}^{-t/\tau} = 9 + [6 - 9] \cdot \mathrm{e}^{-t/(24 \cdot 10^{-6})}\,\text{V}$$

1.7.14 RC. Capacitors in Series Under DC Supply

In the circuit represented in Fig. 1.73, the switch closes at $t = 0$. Calculate the voltages $u_{C1}(t)$ and $u_{C2}(t)$ for $t > 0$, knowing that $u_{C1}(0^-) = 12\,\text{V}$ and $u_{C2}(0^-) = -2\,\text{V}$.

Solution. Before the switch closing, the initial charging voltages of the capacitors is provided by the statement. Except for an impulse-type response, the voltage in each capacitor does not change when the switch is closed, therefore:

$$u_{C1}(0^+) = u_{C1}(0^-) = 12\,\text{V}$$
$$u_{C2}(0^+) = u_{C2}(0^-) = -2\,\text{V}$$

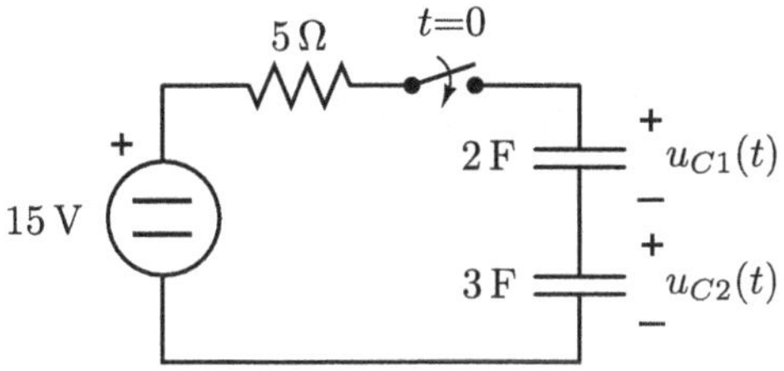

Fig. 1.73 Circuit considered in the problem under study

On the other hand, it is known that the voltage of each capacitor can be expressed as follows:

$$\begin{aligned} u_{C1}(t) &= u_{C1}(0^+) + \frac{1}{C_1}\int_{0^+}^{t} i(\lambda)d\lambda = 12 + u_{C1}^d(t) \\ u_{C2}(t) &= u_{C2}(0^+) + \frac{1}{C_2}\int_{0^+}^{t} i(\lambda)d\lambda = -2 + u_{C2}^d(t) \end{aligned} \tag{1.49}$$

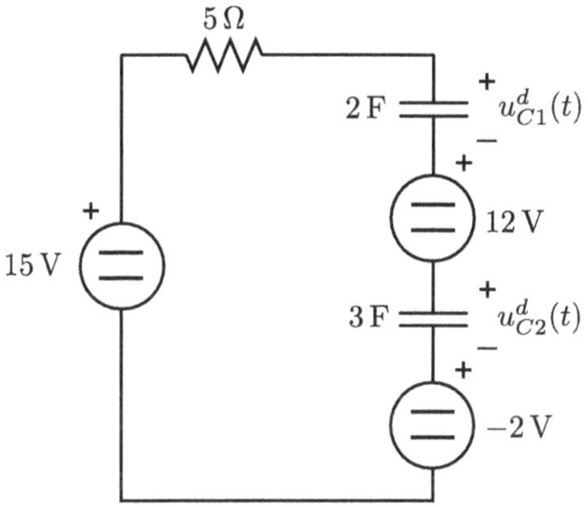

Fig. 1.74 Circuit for $t > 0$

where $u_{C1}^d(t)$ and $u_{C2}^d(t)$ are the voltages of the two discharged capacitors.[1] For this reason, the circuit shown in Fig. 1.74 is obtained, according to which it can be verified that both discharged capacitors and the voltage sources related to the initial conditions can be associated in series, resulting in the circuit depicted in Fig. 1.75, where

$$u_{Ceq}^d(t) = u_{C1}^d(t) + u_{C2}^d(t)$$

and

$$C_{eq} = \frac{C_1 \cdot C_2}{C_1 + C_2} = \frac{2 \cdot 3}{2+3} = \frac{6}{5}\,\mathrm{F}$$

[1] The variable λ has been used to prevent the integration variable from matching the upper limit.

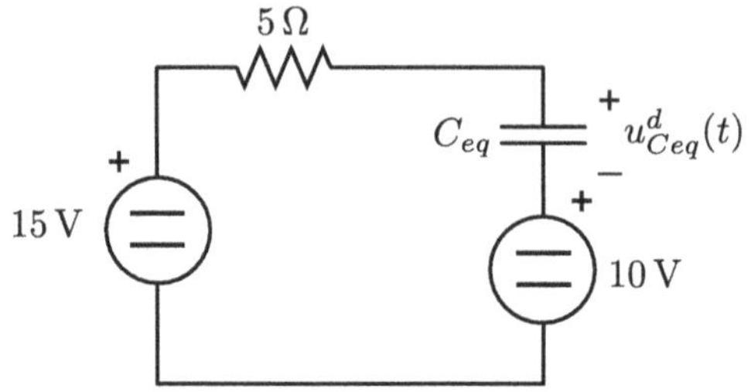

Fig. 1.75 Simplified circuit

It is observed in the resulting circuit in Fig. 1.75 an RC circuit with dc excitation, where the capacitor voltage can be calculated as follows:

$$u^d_{Ceq}(t) = u^{dp}_{Ceq}(t) + \left[u^d_{Ceq}(0^+) - u^{dp}_{Ceq}(0^+)\right] \cdot e^{-t/\tau}$$

Given that the equivalent capacitor is discharged:

$$u^d_{Ceq}(0^+) = 0\,\text{V}$$

The steady-state voltage, $u^{dp}_{Ceq}(t)$, is obtained by solving the steady-state circuit shown in Fig. 1.76, where the capacitor has been replaced by an open circuit.

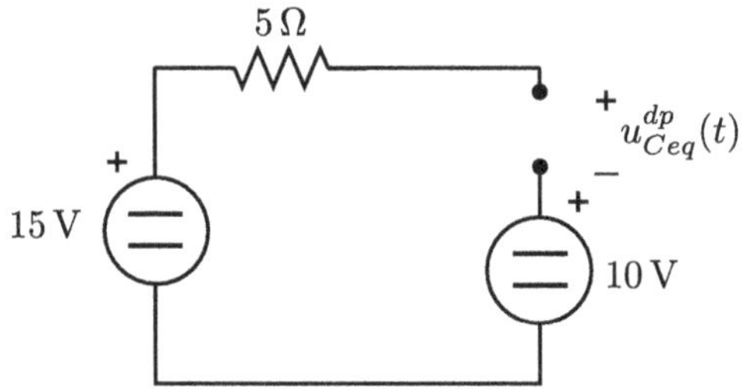

Fig. 1.76 Circuit in dc steady-state regime

From the circuit in Fig. 1.76:

$$u^{dp}_{Ceq}(t) = 15 - 10 = 5\,\text{V} \;\; ; \;\; u^{dp}_{Ceq}(0^+) = 5\,\text{V}$$

Being an RC type circuit, the time constant is calculated as follows:

$$\tau = RC_{eq} = 5 \cdot \frac{6}{5} = 6\,\text{s}$$

Thus:

$$u^d_{Ceq}(t) = u^{dp}_{Ceq}(t) + \left[u^d_{Ceq}(0^+) - u^{dp}_{Ceq}(0^+)\right] \cdot e^{-t/\tau} = 5 + [0 - 5] \cdot e^{-t/6} = 5 \cdot \left[1 - e^{-t/6}\right]\,\text{V}$$

Once the voltage of the discharged equivalent capacitor has been obtained, the voltage of each discharged capacitor is calculated by applying the concept of a capacitive voltage divider:

$$u^d_{C1}(t) = u^d_{Ceq}(t) \cdot \frac{\frac{1}{C_1}}{\frac{1}{C_1} + \frac{1}{C_2}} = u^d_{Ceq}(t) \cdot \frac{C_2}{C_1 + C_2} = u^d_{Ceq}(t) \cdot \frac{3}{5} = 3 \cdot \left[1 - e^{-t/6}\right] \text{V}$$

$$u^d_{C2}(t) = u^d_{Ceq}(t) \cdot \frac{\frac{1}{C_2}}{\frac{1}{C_1} + \frac{1}{C_2}} = u^d_{Ceq}(t) \cdot \frac{C_1}{C_1 + C_2} = u^d_{Ceq}(t) \cdot \frac{2}{5} = 2 \cdot \left[1 - e^{-t/6}\right] \text{V}$$

Finally, considering the relation (1.49), the complete responses of each capacitor voltages are obtained:

$$u_{C1}(t) = 12 + 3 \cdot \left[1 - e^{-t/6}\right] = 15 - 3 \cdot e^{-t/6} \text{V}$$
$$u_{C2}(t) = -2 + 2 \cdot \left[1 - e^{-t/6}\right] = -2 \cdot e^{-t/6} \text{V}$$

The time evolution of $u_{C1}(t)$ and $u_{C2}(t)$ is graphically represented in Fig. 1.77.

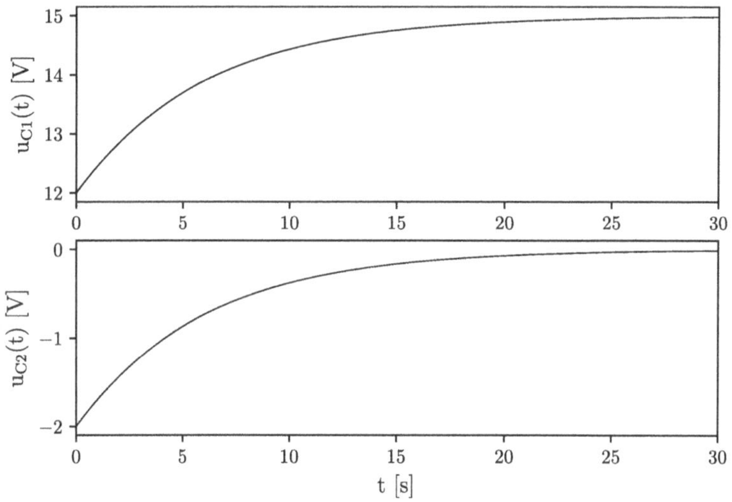

Fig. 1.77 Time evolution of $u_{C1}(t)$ and $u_{C2}(t)$

1.7.15 RL. Inductors in Series Under DC Supply

In the circuit represented in Fig. 1.78, the switch closes at $t = 0$. Calculate the current $i(t)$ for $t > 0$.

Solution. It can be seen in Fig. 1.78 the series connection of two inductors through which, initially, no current flows. Given the current equality, the inductors can be

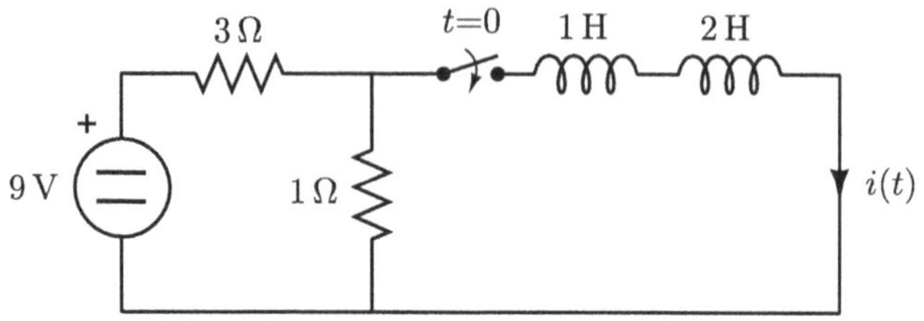

Fig. 1.78 Circuit considered in the problem under study

replaced by a single equivalent one whose initial current will be the same as that of each of them. So that:

$$L_{eq} = L_1 + L_2 = 1 + 2 = 3\,\text{H}$$

This way, the resulting circuit is that shown in Fig. 1.79.

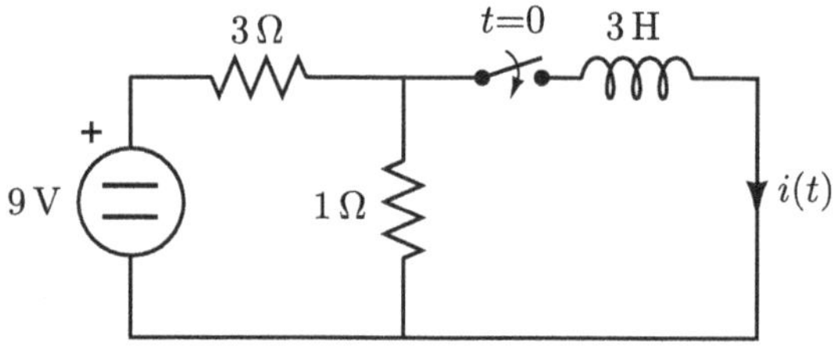

Fig. 1.79 Simplified circuit

Before the switch closing, no current flows through the inductor, so that $i(0^-) = 0\,\text{A}$. Except for an impulse-type response, the current of an inductor does not change when the switch closes, therefore:

$$i(0^+) = i(0^-) = 0\,\text{A}$$

The steady-state value of the current $i(t)$ is subsequently obtained. In this case, once the switch is closed, only a dc-type source of excitation is present in the resulting circuit. For this reason, the dc circuit shown in Fig. 1.80 will be solved, where the inductor has been replaced by a short circuit.

From the circuit in Fig. 1.80:

$$i^p(t) = \frac{9}{3} = 3\,\text{A}$$

Substituting $t = 0^+$:

$$i^p(0^+) = 3\,\text{A}$$

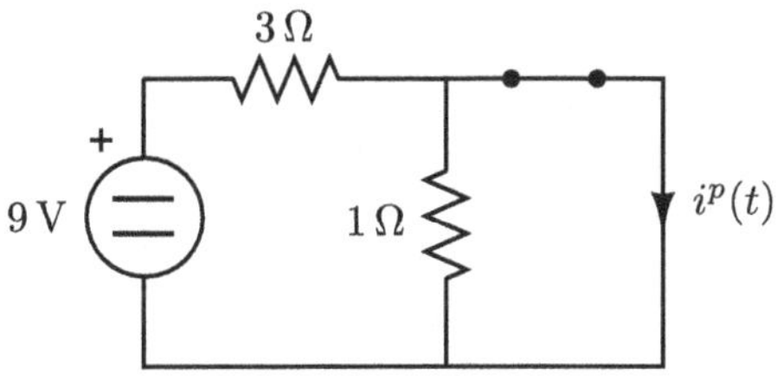

Fig. 1.80 Circuit in dc steady-state regime

Being an RL type circuit, the time constant is calculated as:

$$\tau = \frac{L_{eq}}{R_{eq}}$$

where R_{eq} is the equivalent resistance, from the terminals of the equivalent inductor, corresponding to the parallel connection of the 3 Ω resistance and the 1 Ω resistance, yielding:

$$\tau = \frac{L_{eq}}{R_{eq}} = \frac{3}{\frac{3 \cdot 1}{3+1}} = 4\,\mathrm{s}$$

Finally, the complete response of $i(t)$ for $t > 0$ is consequently derived:

$$i(t) = i^p(t) + \left[i(0^+) - i^p(0^+)\right] \cdot \mathrm{e}^{-t/\tau} = 3 + [0 - 3] \cdot \mathrm{e}^{-t/4} = 3 \cdot (1 - \mathrm{e}^{-t/4})\,\mathrm{A}$$

The time evolution of $i(t)$ has been graphically represented in Fig. 1.81.

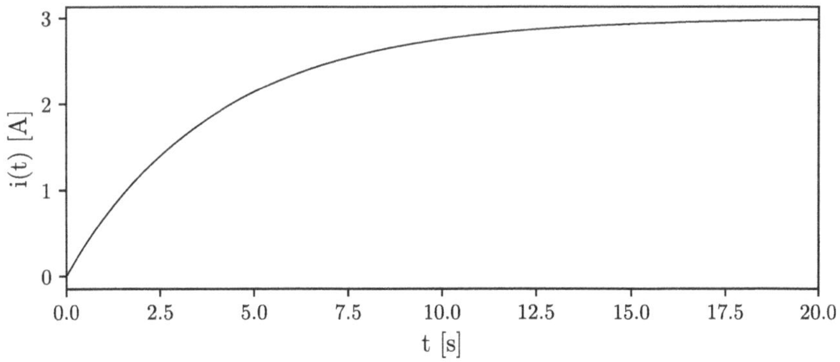

Fig. 1.81 Time evolution of $i(t)$

1.7.16 RL. Inductors in Parallel Under DC Supply

The circuit represented in Fig. 1.82 is in steady state when, at $t = 0$, the switch k1 is closed and switch k2 is opened. Calculate the currents $i_{L1}(t)$ and $i_{L2}(t)$ for $t > 0$.

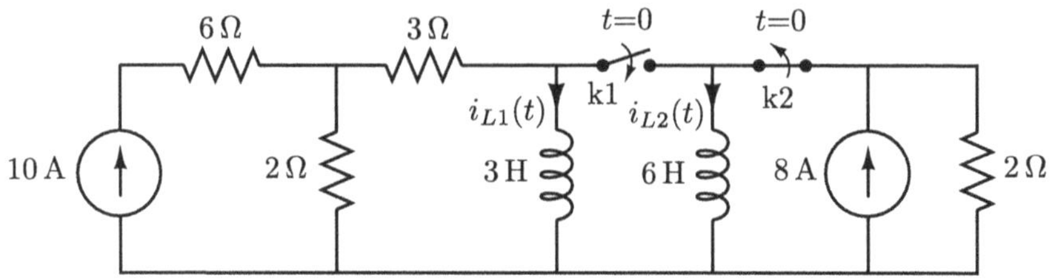

Fig. 1.82 Circuit considered in the problem under study

Solution. Before the switch k2 opens, it can be noticed that the current through inductor 2 is 8 A, meaning:

$$i_{L2}(0^-) = 8\,\text{A}$$

On the other hand, the current through inductor 1 before switch k1 closes ($t = 0^-$) is obtained considering that the circuit is in steady state before the switch closing. In this situation, only a dc-type source of excitation is present, so that the dc circuit shown in Fig. 1.83 will be solved, where inductor 1 has been replaced by a short circuit.

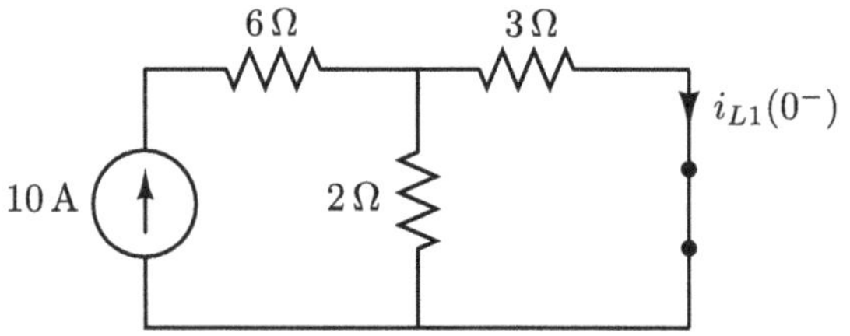

Fig. 1.83 Circuit at $t = 0^-$

From the circuit in Fig. 1.83, the current through inductor 1 is obtained at $t = 0^-$:

$$i_{L1}(0^-) = 10 \cdot \frac{\frac{1}{3}}{\frac{1}{2} + \frac{1}{3}} = 4\,\text{A}$$

Except for an impulse-type response, which is not the case, the currents through the inductors do not change when the switch is closed, therefore:

$$i_{L1}(0^+) = i_{L1}(0^-) = 4\,\text{A}$$
$$i_{L2}(0^+) = i_{L2}(0^-) = 8\,\text{A}$$

On the other hand, it is known that the current of each inductor can be expressed as follows

$$
\begin{aligned}
i_{L1}(t) &= i_{L1}(0^+) + \frac{1}{L_1}\int_{0^+}^{t} u(\lambda)d\lambda = 4 + i^d_{L1}(t) \\
i_{L2}(t) &= i_{L2}(0^+) + \frac{1}{L_2}\int_{0^+}^{t} u(\lambda)d\lambda = 8 + i^d_{L2}(t)
\end{aligned}
\tag{1.50}
$$

where $i^d_{L1}(t)$ and $i^d_{L2}(t)$ are the currents of the discharged inductors.[2] The resulting circuit for $t > 0$ is shown in Fig. 1.84.

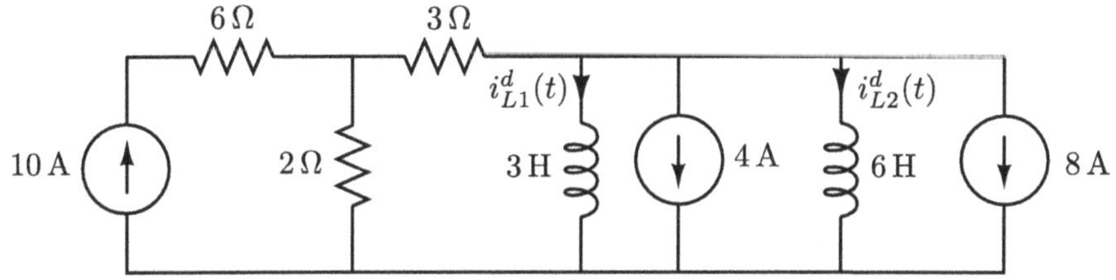

Fig. 1.84 Circuit for $t > 0$

From the circuit represented Fig. 1.84, it can be verified that the discharged inductors and the corresponding current sources can be associated in parallel, resulting in the circuit depicted in Fig. 1.85, where

$$i^d_{Leq}(t) = i^d_{L1}(t) + i^d_{L1}(t)$$

and

$$L_{eq} = \frac{L_1 \cdot L_2}{L_1 + L_2} = \frac{3 \cdot 6}{3 + 6} = 2\,\text{H}$$

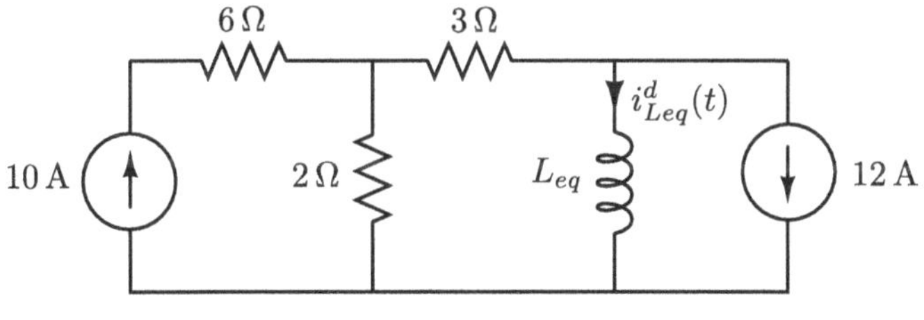

Fig. 1.85 Simplified circuit

It can be noticed in the resulting circuit in Fig. 1.85 that it is an RL circuit under dc supply, where the equivalent inductor current can be calculated using the following expression:

$$i^d_{Leq}(t) = i^{dp}_{Leq}(t) + \left[i^d_{Leq}(0^+) - i^{dp}_{Leq}(0^+)\right] \cdot \text{e}^{-t/\tau}$$

---

[2] The variable λ has been used to prevent the integration variable from matching the upper limit.

Given that the equivalent inductor is discharged:

$$i_{Leq}^{d}(0^+) = 0\,\mathrm{A}$$

The steady-state current, $i_{Leq}^{dp}(t)$, is obtained by solving the steady-state circuit shown in Fig. 1.86, where the inductor has been replaced by a short circuit.

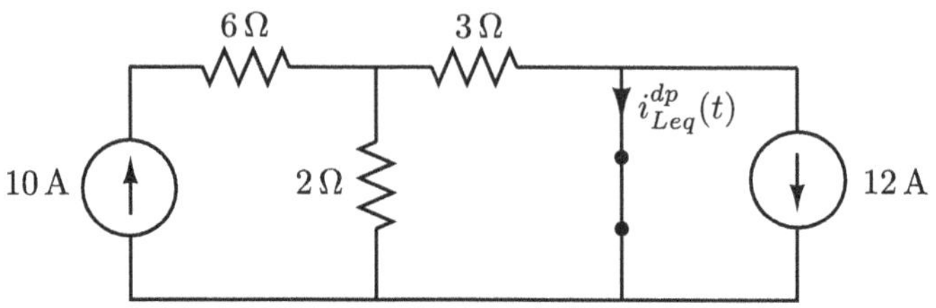

Fig. 1.86 Circuit in dc steady-state regime

From the circuit in Fig. 1.86:

$$i_{Leq}^{dp}(t) = -12 + 10 \cdot \frac{\frac{1}{3}}{\frac{1}{2} + \frac{1}{3}} = -8\,\mathrm{A}$$

Substituting $t = 0^+$:

$$i_{Leq}^{dp}(0^+) = -8\,\mathrm{A}$$

Being an RL type circuit, the time constant is obtained as:

$$\tau = \frac{L_{eq}}{R_{eq}}$$

where R_{eq} is the equivalent resistance from the inductor terminals, corresponding to the series connection of the 3 Ω resistance and the 2 Ω resistance, yielding:

$$\tau = \frac{L_{eq}}{R_{eq}} = \frac{2}{3+2} = \frac{2}{5}\,\mathrm{s}$$

Thus:

$$\begin{aligned} i_{Leq}^{d}(t) &= i_{Leq}^{dp}(t) + \left[i_{Leq}^{d}(0^+) - i_{Leq}^{dp}(0^+) \right] \cdot \mathrm{e}^{-t/\tau} \\ &= -8 + [0 + 8] \cdot \mathrm{e}^{-5t/2} = 8 \cdot \left[-1 + \mathrm{e}^{-5t/2} \right] \mathrm{A} \end{aligned}$$

Once the current of the discharged equivalent inductor has been obtained, the currents of each discharged inductor are calculated using the concept of inductive current divider:

$$i_{L1}^{d}(t) = i_{Leq}^{d}(t) \cdot \frac{\frac{1}{L_1}}{\frac{1}{L_1}+\frac{1}{L_2}} = i_{Leq}^{d}(t) \cdot \frac{L_2}{L_1+L_2} = i_{Leq}^{d}(t) \cdot \frac{6}{9} = \frac{48}{9} \cdot \left[-1+\mathrm{e}^{-5t/2}\right] \text{A}$$

$$i_{L2}^{d}(t) = i_{Leq}^{d}(t) \cdot \frac{\frac{1}{L_2}}{\frac{1}{L_1}+\frac{1}{L_2}} = i_{Leq}^{d}(t) \cdot \frac{L_1}{L_1+L_2} = i_{Leq}^{d}(t) \cdot \frac{3}{9} = \frac{24}{9} \cdot \left[-1+\mathrm{e}^{-5t/2}\right] \text{A}$$

Finally, using the relation (1.50), the complete responses of the each inductor currents are obtained:

$$i_{L1}(t) = 4 + \frac{48}{9} \cdot \left[-1+\mathrm{e}^{-5t/2}\right] \text{A}$$

$$i_{L2}(t) = 8 + \frac{24}{9} \cdot \left[-1+\mathrm{e}^{-5t/2}\right] \text{A}$$

The time evolution of $i_{L1}(t)$ and $i_{L2}(t)$ is graphically represented in Fig. 1.87.

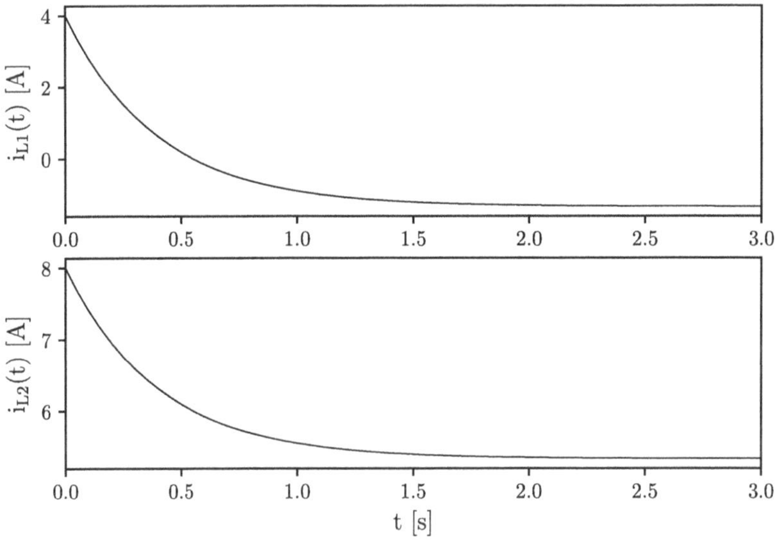

Fig. 1.87 Time evolution of $i_{L1}(t)$ and $i_{L2}(t)$

1.7.17 RL. Concatenated Transients Under DC Supply

In the circuit represented in Fig. 1.88, the switch is in the position 0. At $t = 0$, the switch changes to position 1 and, at $t = 500\,\mu$s, changes to position 2. Calculate $i(t)$ for $t > 0$.

Solution. It can be noticed from the circuit in Fig. 1.88 that, depending on the state of the switch, different circuits are formed and, consequently, they will have to be solved separately. First, the circuit in the time interval $0 < t < 500\,\mu$s will be analyzed, being the second circuit that in the interval $t > 500\,\mu$s.

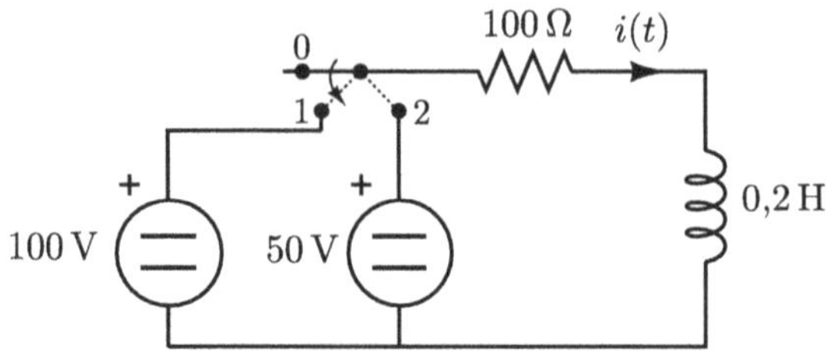

Fig. 1.88 Circuit considered in the problem under study

Circuit analysis for $0 < t < 500\,\mu s$
When the switch is in the position 1, the resulting circuit is shown in Fig. 1.89.

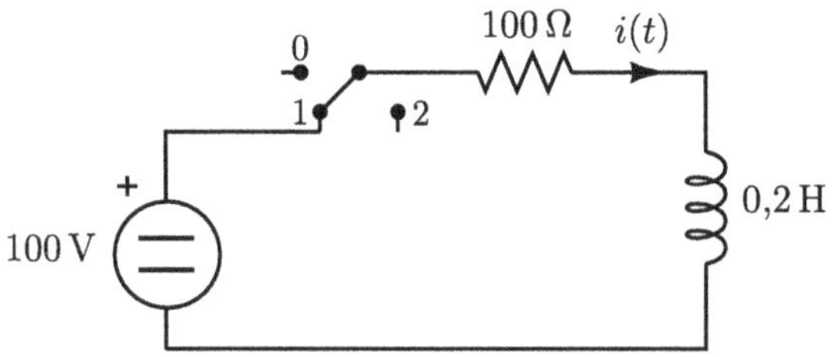

Fig. 1.89 Circuit for $0 < t < 500\,\mu s$

With the switch in the position 0, no current flows through the inductor, so that $i(0^-) = 0\,\text{A}$. Except for an impulse-type response, which is not the case, the inductor current immediately after the switch goes to the position 1 does not change with respect to the previous one, therefore:

$$i(0^+) = i(0^-) = 0\,\text{A}$$

The steady-state current, $i^p(t)$, will be subsequently obtained. In this case, only a dc-type source of excitation is present, so that the dc circuit shown in Fig. 1.90 will be solved, where the inductor has been replaced by a short circuit.

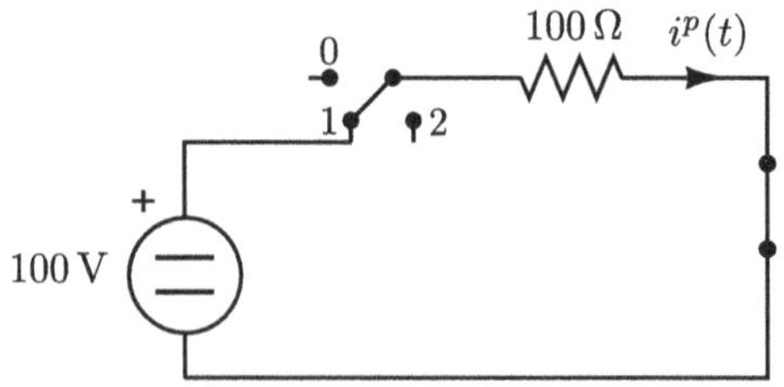

Fig. 1.90 Circuit in dc steady-state regime

From the circuit in Fig. 1.90:

$$i^p(t) = \frac{100}{100} = 1\,\text{A}$$

Substituting $t = 0^+$:

$$i^p(0^+) = 1\,\text{A}$$

Being an RL type circuit, the time constant is calculated as:

$$\tau = \frac{L}{R} = \frac{0{,}2}{100} = \frac{1}{500}\,\text{s}$$

Consequently, the complete response of $i(t)$ for $0 < t < 500\,\mu\text{s}$ is derived:

$$i(t) = i^p(t) + \left[i(0^+) - i^p(0^+)\right] \cdot \text{e}^{-t/\tau} = 1 + [0 - 1] \cdot \text{e}^{-500t} = 1 - \text{e}^{-500t}\,\text{A}$$

Before the switch changes to the position 2 ($t = 500\,\mu\text{s}$), the inductor current is:

$$i(500\,\mu\text{s}^-) = 1 - \text{e}^{-500 \cdot 500 \cdot 10^{-6}} \approx 0{,}221\,\text{A}$$

This value will be used to solve the next transient.

Circuit analysis for $t > 500\,\mu\text{s}$

When the switch changes to the position 2, the resulting circuit is shown in Fig. 1.91.

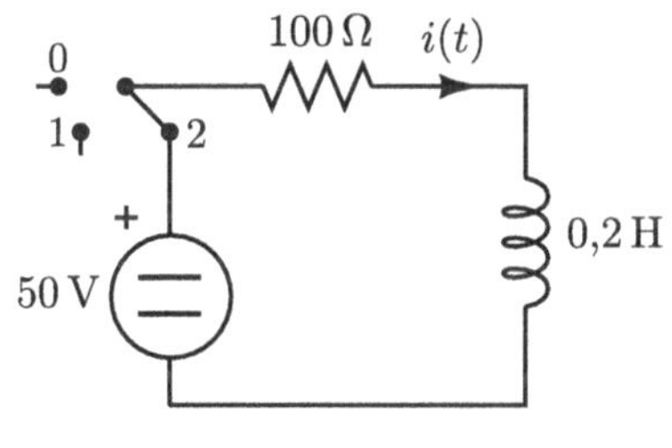

Fig. 1.91 Circuit for $t > 500\,\mu\text{s}$

Except for an impulse-type response, which is not the case, the inductor current, immediately after the switch moves to the position 2, has the same value as that immediately before changing to the position 2, so that:

$$i(500\,\mu\text{s}^+) = i(500\,\mu\text{s}^-) = 0{,}221\,\text{A}$$

The steady-state current, $i^p(t)$, will be subsequently obtained. In this case, only a dc-type source of excitation is present, so that the dc circuit shown in Fig. 1.92 will be solved, where the inductor has been replaced by a short circuit.

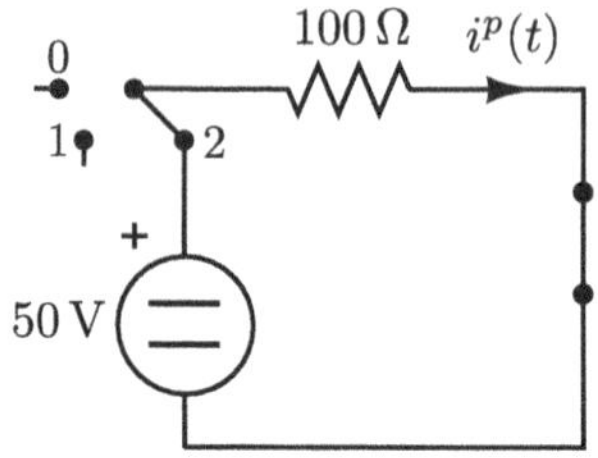

Fig. 1.92 Circuit in DC steady-state regime

From the circuit in Fig. 1.90:

$$i^p(t) = \frac{50}{100} = 0{,}5\,\text{A}$$

Substituting $t = 500\,\mu\text{s}^+$:

$$i^p(500\,\mu\text{s}^+) = 0{,}5\,\text{A}$$

The time constant is the same as in the previous case since the values of R and L have not changed. So that:

$$\tau = \frac{L}{R} = \frac{0{,}2}{100} = \frac{1}{500}\,\text{s}$$

Finally, the complete response of $i(t)$ for $t > 500\,\mu\text{s}$ is consequently derived:

$$\begin{aligned} i(t) &= i^p(t) + \left[i(500\,\mu\text{s}^+) - i^p(500\,\mu\text{s}^+)\right] \cdot \text{e}^{-(t-500\cdot 10^{-6})/\tau} \\ &= 0{,}5 + [0{,}221 - 0{,}5] \cdot \text{e}^{-500(t-500\cdot 10^{-6})}\,\text{A} \end{aligned}$$

The time evolution of $i(t)$ has been graphically represented in Fig. 1.93.

1.7.18 RC. Concatenated Transients Without Excitation Sources

The circuit represented in Fig. 1.94 is in steady state when the switch goes from position 0 to position 1 at $t = 0$, while the switch changes to position 2 at $t = 10\,\text{ms}$. Calculate $i(t)$ for $t > 0$.

Solution. It can be noticed from the circuit in Fig. 1.94 that, depending on the state the switch, different circuits are formed and, consequently, they will be analyzed separately. First, the circuit corresponding to the switch in position 0 will be analyzed.

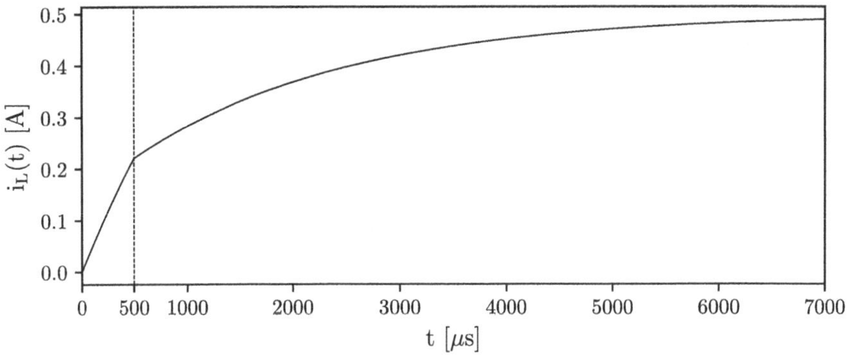

Fig. 1.93 Time evolution of $i(t)$

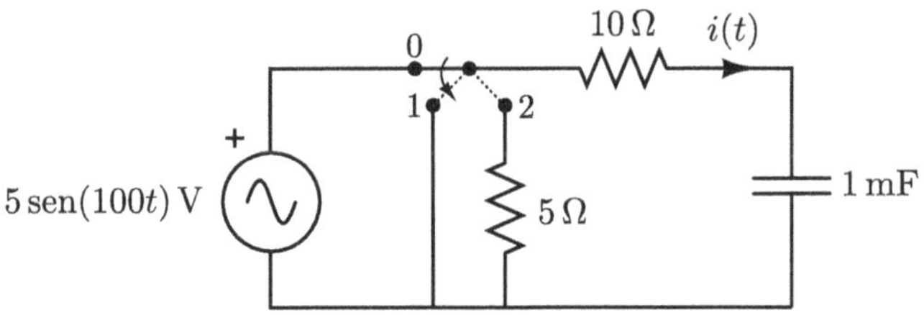

Fig. 1.94 Circuit considered in the problem under study

According to the problem statement, the circuit is in steady state when the switch is in the position 0. In this case, only an ac-type source of excitation is present, so that the circuit in the phasor domain shown in Fig. 1.95 will be solved in steady-state ac regime. The objective consists of obtaining the capacitor voltage at the instant before the switch changes to the position 1, that is, $u_C(0^-)$.

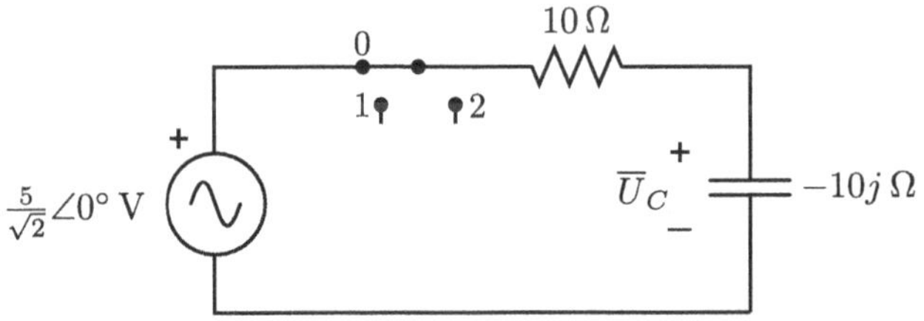

Fig. 1.95 Circuit in the phasor domain

From the circuit in Fig. 1.95:

$$\overline{U}_C = \frac{5}{\sqrt{2}}\angle 0° \cdot \frac{-10j}{10-10j} = 2{,}5\angle -45° \text{ V}$$

Once the capacitor voltage in the phasor domain has been calculated, the corresponding expression in the time domain is obtained:

$$u_C(t) = 2{,}5\sqrt{2}\sin(100t - 45°)\,\mathrm{V}$$

whose value for $t = 0^-$ is:

$$u_C(0^-) = 2{,}5\sqrt{2}\sin(-45°) = -2{,}5\,\mathrm{V}$$

The circuit is subsequently analyzed in the time intervals $0 < t < 10\,\mathrm{ms}$, and $t > 10\,\mathrm{ms}$.

Circuit analysis for $0 < t < 10\,\mathrm{ms}$

When the switch changes to position 1, the resulting circuit is shown in Fig. 1.96. Although the requested variable is the current, the capacitor voltage will be also calculated, since it is required for the resolution of the transient beginning at 10 ms.

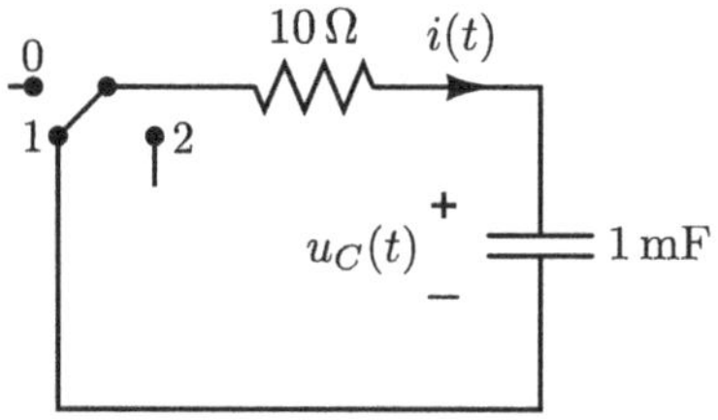

Fig. 1.96 Circuit for $0 < t < 10\,\mathrm{ms}$

To obtain the initial condition, the circuit at instant $t = 0^+$will be used, which is obtained from that in Fig. 1.96, substituting the capacitor for a voltage source of value $u_C(0^+)$. Except for an impulse-type response:

$$u_C(0^+) = u_C(0^-) = -2{,}5\,\mathrm{V}$$

The circuit at time $t = 0^+$ is shown in Fig. 1.97.

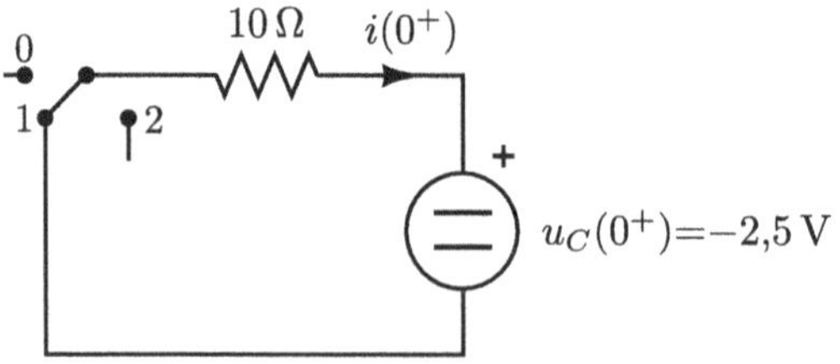

Fig. 1.97 Circuit at $t = 0^+$

From the circuit in Fig. 1.97, the value of $i(0^+)$ can be easily obtained:

$$i(0^+) = \frac{2{,}5}{10} = 0{,}25\,\text{A}$$

The steady-state value of the current $i(t)$ is subsequently obtained. In this case, there is no excitation sources in the resulting circuit, only the capacitor connected to the resistor. Consequently:

$$i^p(t) = 0\,\text{A} \;\;;\;\; u_C^p(t) = 0\,\text{V}$$

Substituting $t = 0^+$:

$$i^p(0^+) = 0\,\text{A} \;\;;\;\; u_C^p(0^+) = 0\,\text{V}$$

Being an RC type circuit, the time constant is calculated as:

$$\tau = RC = 10 \cdot 1 \cdot 10^{-3} = \frac{1}{100}\,\text{s}$$

The complete response of $i(t)$ and $u_C(t)$ for $0 < t < 10$ ms are consequently derived:

$$i(t) = i^p(t) + \left[i(0^+) - i^p(0^+)\right] \cdot \text{e}^{-t/\tau} = 0 + [0{,}25 - 0] \cdot \text{e}^{-100t} = 0{,}25 \cdot \text{e}^{-100t}\,\text{A}$$
$$u_C(t) = u_C^p(t) + \left[u_C(0^+) - u_C^p(0^+)\right] \cdot \text{e}^{-t/\tau} = 0 + [-2{,}5 - 0] \cdot \text{e}^{-100t} = -2{,}5 \cdot \text{e}^{-100t}\,\text{V}$$

Before the switch changes to position 2 at $t = 10\,\text{ms}$, the capacitor voltage is:

$$u_C(10\,\text{ms}^-) = -2{,}5 \cdot \text{e}^{-100 \cdot 10 \cdot 10^{-3}} \approx -0{,}92\,\text{V}$$

Circuit analysis for $t > 10\,\text{ms}$

When the switch changes to position 2, the resulting circuit is shown in Fig. 1.98.

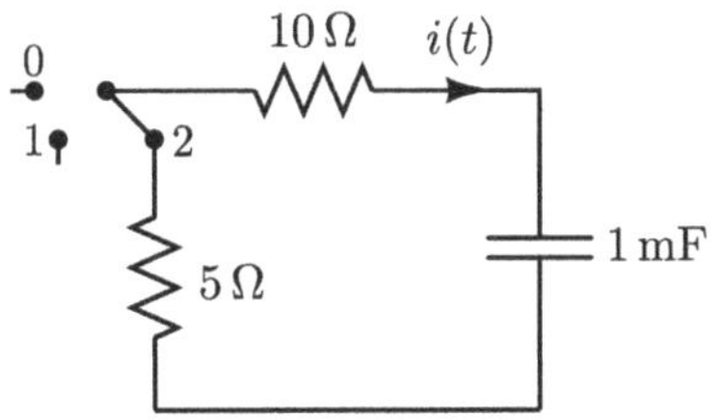

Fig. 1.98 Circuit for $t > 10\,\text{ms}$

To obtain the initial condition, the circuit at instant $t = 10\,\text{ms}^+$ will be used, which is obtained from that in Fig. 1.98 substituting the capacitor for a voltage source of value $u_C(10\,\text{ms}^+)$. Except for an impulse-type response:

$$u_C(10\,\text{ms}^+) = u_C(10\,\text{ms}^-) = -0{,}92\,\text{V}$$

The circuit at time $t = 10\,\text{ms}^+$ is shown in Fig. 1.99, from which it can be observed that:

$$i(10\,\text{ms}^+) = \frac{0{,}92}{5+10} \approx 0{,}061\,\text{A}$$

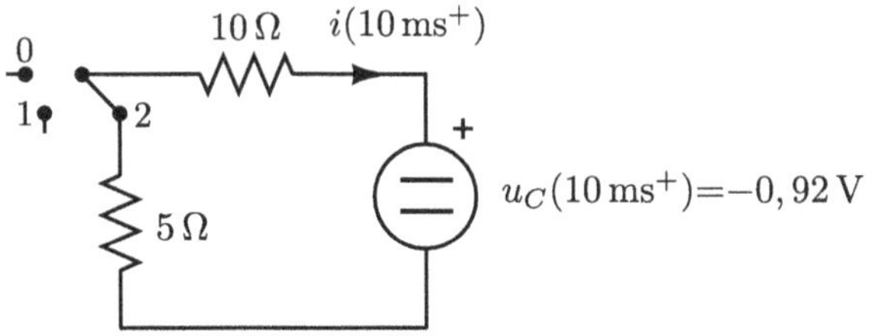

Fig. 1.99 Circuit at $t = 10\,\text{ms}^+$

As in the previous case, there is no excitation sources in the resulting circuit, remaining only the capacitor connected to the two resistors. Consequently:

$$i^p(t) = 0\,\text{A} \ ; \ i^p(10\,\text{ms}^+) = 0\,\text{A}$$

Being an RC circuit, the time constant (for $t > 10\,\text{ms}$) is calculated as:

$$\tau = R_{eq}C$$

where R_{eq} is the equivalent resistance from the capacitor terminals, corresponding to the series connection of the $5\,\Omega$ resistor and the $10\,\Omega$ resistor, yielding:

$$\tau = R_{eq}C = (5+10)\cdot 1\cdot 10^{-3} = 15\cdot 10^{-3}\,\text{s}$$

Finally, the complete response of $i(t)$ for $t > 10\,\text{ms}$ is obtained:

$$\begin{aligned} i(t) &= i^p(t) + \left[i(10\,\text{ms}^+) - i^p(10\,\text{ms}^+)\right]\cdot \text{e}^{-(t-10\cdot 10^{-3})/\tau} \\ &= 0 + [0{,}061 - 0]\cdot \text{e}^{-(t-10\cdot 10^{-3})/(15\cdot 10^{-3})}\,\text{A} \end{aligned}$$

The time evolution of $i(t)$ has been graphically represented in Fig. 1.100.

1.7.19 Capacitors in Parallel. Impulse Response

In the circuit represented in Fig. 1.101 it is known that $u_{C1}(0^-) = 5\,\text{V}$, $u_{C2}(0^-) = 0\,\text{V}$, $C_1 = 1\,\text{F}$ and $C_2 = 0{,}5\,\text{F}$. At $t = 0$, the switch closes. Calculate the voltage of the capacitors and the current $i(t)$ for $t > 0$.

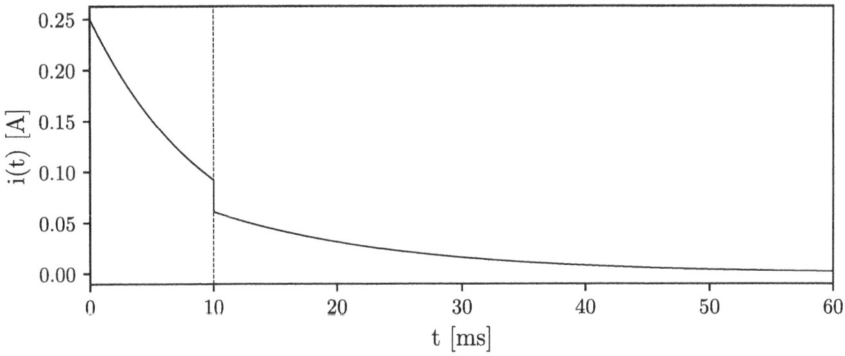

Fig. 1.100 Time evolution of $i(t)$

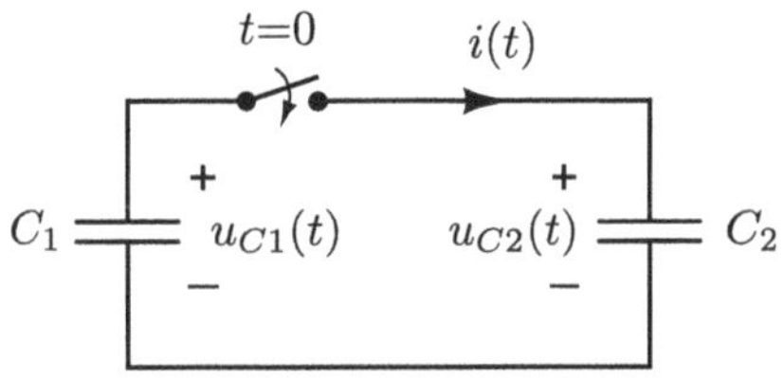

Fig. 1.101 Circuit considered in the problem under study

Solution. According to the problem statement, before the switch closes, the capacitors are charged with a different voltage. Specifically, one of them has a voltage of 5 V and the other one a voltage of 0 V. When the switch is closed, both capacitors are in parallel. In the instant immediately after the closure of said switch, the capacitors are subjected to the same voltage at the expense of a current impulse. In this instantaneous process, the total charge stored in both capacitors does not change, which allows obtaining the common voltage as:

$$u_{C1}(0^+) = u_{C2}(0^+) = \frac{C_1 \cdot u_{C1}(0^-) + C_2 \cdot u_{C2}(0^-)}{C_1 + C_2} = \frac{1 \cdot 5 + 0{,}5 \cdot 0}{1 + 0{,}5} = \frac{10}{3}\,\mathrm{V}$$

In this particular case, for $t > 0$, the voltage of both capacitors remains constant (Fig. 1.102):

$$u_{C1}(t) = u_{C2}(t) = \frac{10}{3}\,\mathrm{V}$$

On the other hand, according to the polarity references shown in Fig. 1.101, the relationship between the voltage and the current of the capacitor 2 is that described below:

$$i(t) = C_2 \cdot \frac{du_{C2}(t)}{dt}$$

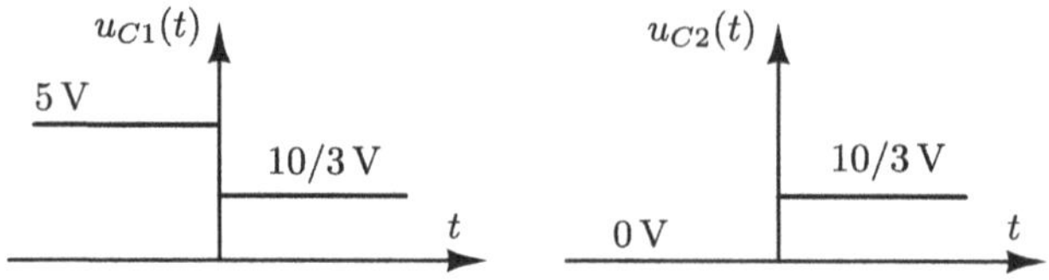

Fig. 1.102 Voltage of both capacitors

This relationship can be expressed as follows:

$$i(t)dt = C_2 \cdot du_{C2}(t)$$

Integrating the previous expression between the infinitesimal limits of 0^+ and 0^-, when the voltage discontinuity occurs in both capacitors, the following result is obtained:

$$\int_{0^-}^{0^+} i(t)dt = \int_{0^-}^{0^+} C_2 \cdot du_{C2}(t) = C_2 \cdot \left[u_{C2}(0^+) - u_{C2}(0^-)\right] = 0{,}5 \cdot \left[\frac{10}{3} - 0\right] = \frac{5}{3}\,C$$

Using the definition of the Dirac delta function

$$\int_{0^-}^{0^+} \delta(t)dt = 1$$

the complete response of the current $i(t)$ for $t > 0$ can be obtained as:

$$i(t) = \frac{5}{3} \cdot \delta(t)\,\text{A}$$

1.7.20 Capacitors in Parallel. Impulse Response

The circuit represented in Fig. 1.103 is in steady state when, at $t = 0$, the switches change their position. Given that $C_1 = 1\,\text{F}$ and $C_2 = 2\,\text{F}$, calculate the voltage of the capacitors and the current $i(t)$ for $t > 0$.

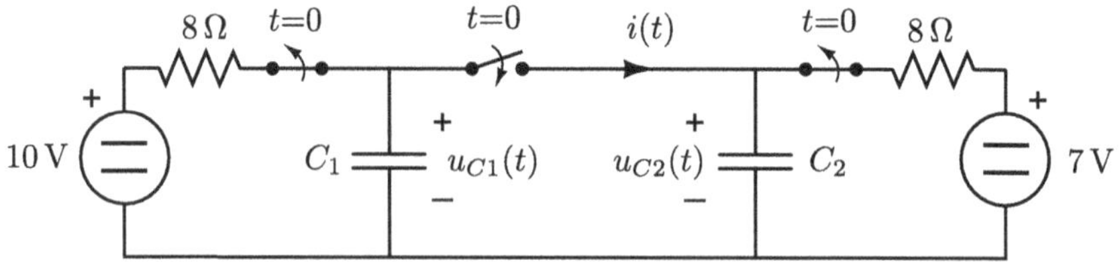

Fig. 1.103 Circuit considered in the problem under study

Solution. First, the voltage in each of the capacitors will be obtained before the switches change their position, that is, at instant $t = 0^-$. To do this, it will be considered that the circuit is in steady-state regime before the position changing. Since, in this situation, only dc-type sources are present, the dc circuit shown in Fig. 1.104 will be solved, where each capacitor has been replaced by an open circuit.

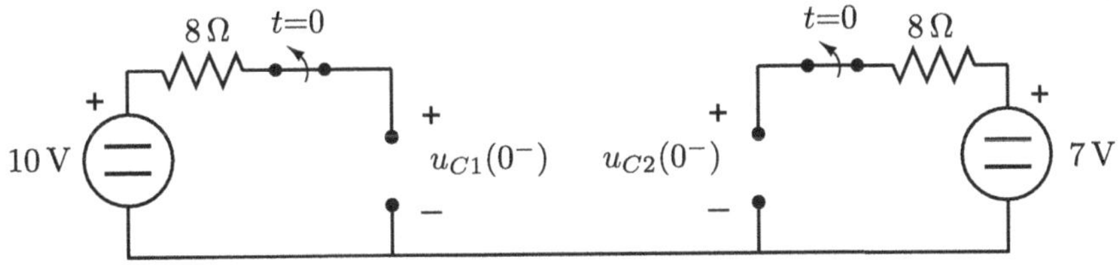

Fig. 1.104 Circuit at $t = 0^-$

From the circuit in Fig. 1.104:

$$u_{C1}(0^-) = 10\,\text{V} \;\; ; \;\; u_{C2}(0^-) = 7\,\text{V}$$

Before the switches change their position, the capacitors are charged to different voltages. When this change occurs, both capacitors are in parallel, as shown in Fig. 1.105. Immediately after the closure of the switch, the capacitors are subjected to the same voltage, at the expense of a current impulse. In this instantaneous process, the total charge stored in both capacitors does not change, which allows obtaining the common voltage as:

$$u_{C1}(0^+) = u_{C2}(0^+) = \frac{C_1 \cdot u_{C1}(0^-) + C_2 \cdot u_{C2}(0^-)}{C_1 + C_2} = \frac{1 \cdot 10 + 2 \cdot 7}{1 + 2} = \frac{24}{3} = 8\,\text{V}$$

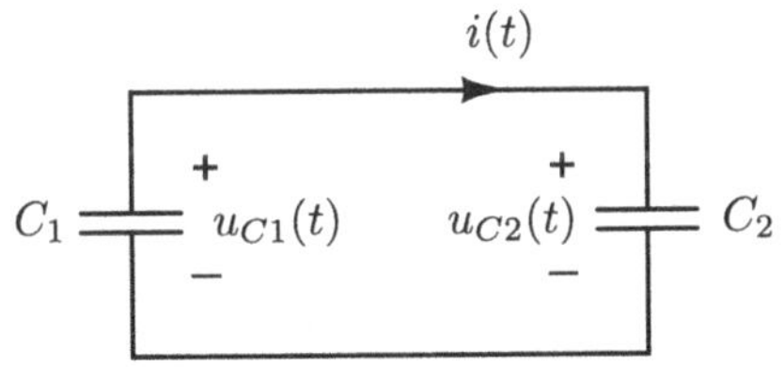

Fig. 1.105 Circuit for $t > 0$

In this particular case, for $t > 0$, the voltage of both capacitors remains constant (Fig. 1.106):

$$u_{C1}(t) = u_{C2}(t) = 8\,\text{V}$$

On the other hand, according to the polarity references shown in Fig. 1.105, the relationship between voltage and current of the capacitor 2 is:

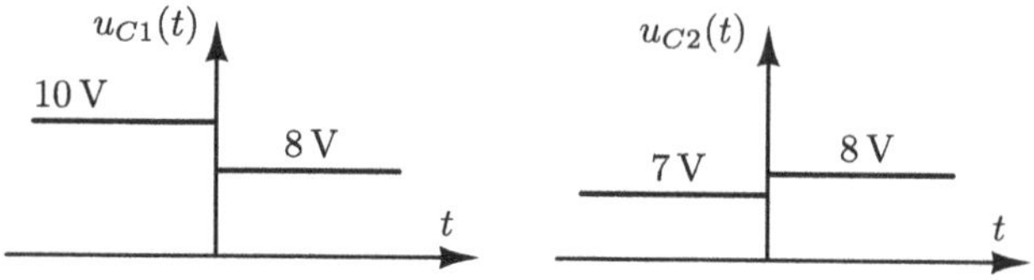

Fig. 1.106 Voltage of both capacitors

$$i(t) = C_2 \cdot \frac{du_{C2}(t)}{dt}$$

This relationship can be expressed as follows:

$$i(t)dt = C_2 \cdot du_{C2}(t)$$

Integrating the previous expression between the infinitesimal limits of 0^+ and 0^-, when the voltage discontinuity occurs in both capacitors, results the expression below:

$$\int_{0^-}^{0^+} i(t)dt = \int_{0^-}^{0^+} C_2 \cdot du_{C2}(t) = C_2 \cdot \left[u_{C2}(0^+) - u_{C2}(0^-)\right] = 2 \cdot [8 - 7] = 2\,C$$

Using the definition of the Dirac delta function, the complete response of $i(t)$ is obtained for $t > 0$:

$$i(t) = 2 \cdot \delta(t)\,\text{A}$$

1.7.21 RC. Impulse Response Under DC Supply

The circuit represented in Fig. 1.107 is in steady state when the switch is closed at $t = 0$. Given that $C_1 = 1\,\text{F}$ and $C_2 = 2\,\text{F}$, calculate the voltage of the capacitors for $t > 0$.

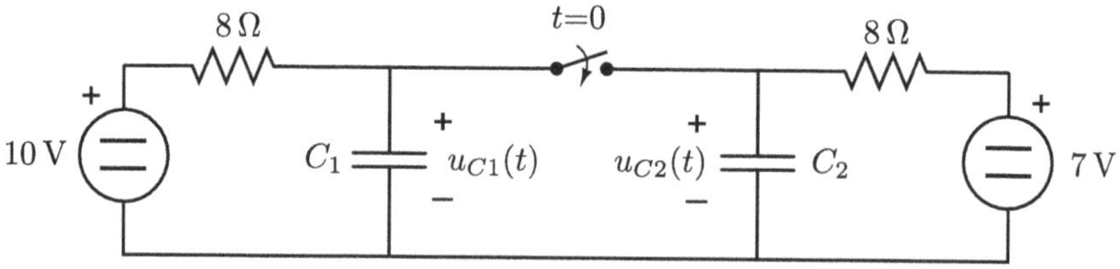

Fig. 1.107 Circuit considered in the problem under study

Solution. First, the voltage of each capacitor will be obtained before the switch closes, that is, at instant $t = 0^-$. For this purpose, it will be considered that the circuit is in steady-state regime before the switch closes. Since, in this situation, only dc-type excitation sources are present, the dc circuit shown in Fig. 1.108 will be solved, where each capacitor has been replaced by an open circuit.

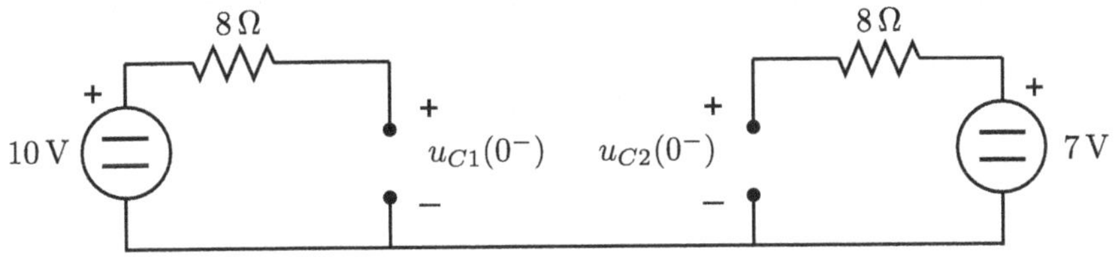

Fig. 1.108 Circuit at $t = 0^-$

From the circuit in Fig. 1.108:

$$u_{C1}(0^-) = 10\,\text{V} \ ; \ \ u_{C2}(0^-) = 7\,\text{V}$$

Before the switch closes, the charging voltages of each capacitor are different. When the switch is closed, both capacitors are connected in parallel as shown in Fig. 1.109.

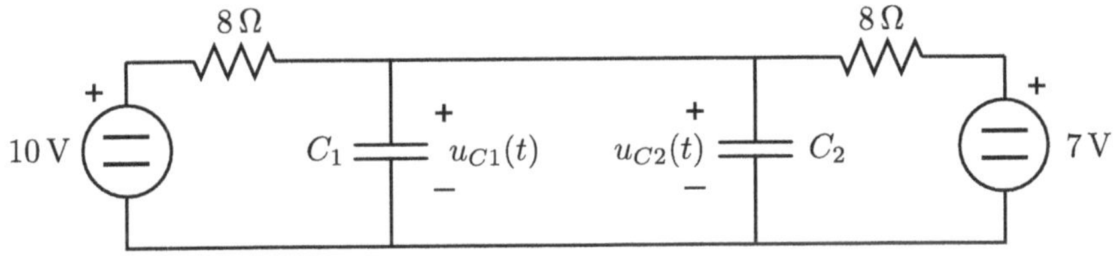

Fig. 1.109 Circuit for $t > 0$

Immediately after the closure of said switch, the capacitors are subjected to the same voltage at the expense of a current impulse. In this instantaneous process, the total charge stored in both capacitors does not change, yielding the following expression for the common voltage:

$$u_{C1}(0^+) = u_{C2}(0^+) = \frac{C_1 \cdot u_{C1}(0^-) + C_2 \cdot u_{C2}(0^-)}{C_1 + C_2} = \frac{1 \cdot 10 + 2 \cdot 7}{1 + 2} = \frac{24}{3} = 8\,\text{V}$$

At instant $t = 0^+$, the capacitors are connected in parallel with an initial voltage of 8 V. In this situation, both capacitors can be replaced by a single equivalent capacitor with

$$C_{eq} = C_1 + C_2 = 1 + 2 = 3\,\text{F}$$

and whose initial voltage is 8 V. This way, the circuit is that shown in Fig. 1.110, where

$$u(0^+) = 8\,\text{V}$$

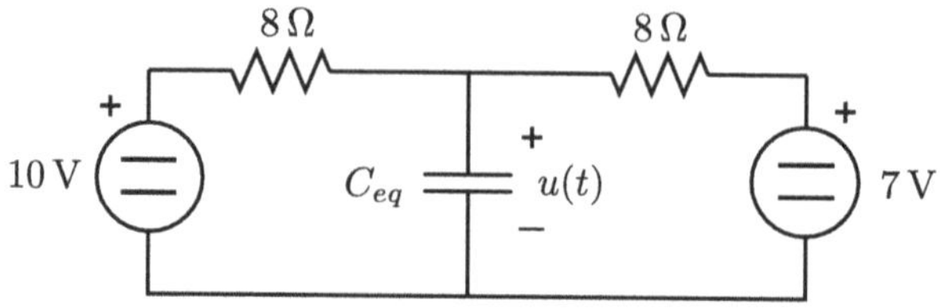

Fig. 1.110 Circuit for $t > 0$

The steady-state value of the voltage $u(t)$ is subsequently obtained. Given that only dc-type sources are present, the dc circuit shown in Fig. 1.111 will be solved, where the capacitor has been replaced by an open circuit.

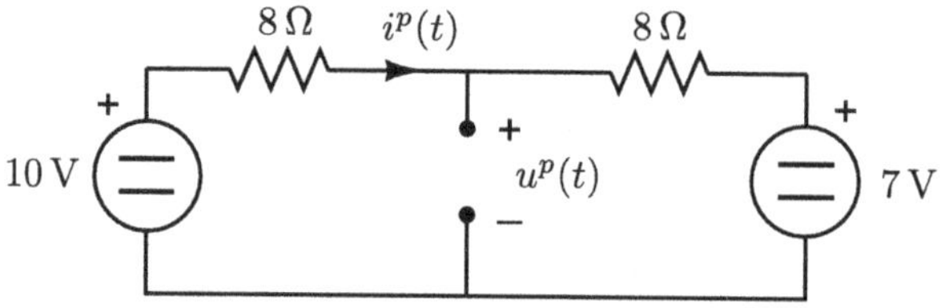

Fig. 1.111 Circuit in dc steady-state regime

From the circuit in Fig. 1.111:

$$i^p(t) = \frac{10-7}{8+8} = \frac{3}{16}\,\mathrm{A}$$

So that:

$$u^p(t) = 10 - 8 \cdot i^p(t) = 10 - 8 \cdot \frac{3}{16} = 8{,}5\,\mathrm{V}$$

Substituting $t = 0^+$:

$$u^p(0^+) = 8{,}5\,\mathrm{V}$$

Being an RC circuit, the time constant is calculated as:

$$\tau = R_{eq} C_{eq}$$

where R_{eq} is the equivalent resistance from the capacitor terminals, corresponding to the parallel connection of both 8 Ω resistors, yielding:

$$\tau = R_{eq} C_{eq} = \frac{8 \cdot 8}{8+8} \cdot 3 = 12\,\mathrm{s}$$

Finally, the complete response of $u(t)$ for $t > 0$ is obtained:

$$u(t) = u^p(t) + \left[u(0^+) - u^p(0^+)\right] \cdot \mathrm{e}^{-t/\tau} = 8{,}5 + [8 - 8{,}5] \cdot \mathrm{e}^{-t/12}\,\mathrm{V}$$

Obviously, for $t > 0$, the voltages of both capacitors are:

$$u_{C1}(t) = u_{C2}(t) = u(t) = 8{,}5 + [8 - 8{,}5] \cdot \mathrm{e}^{-t/12}\,\mathrm{V}$$

The time evolution of $u_{C1}(t)$ and $u_{C2}(t)$ has been graphically represented in Fig. 1.112.

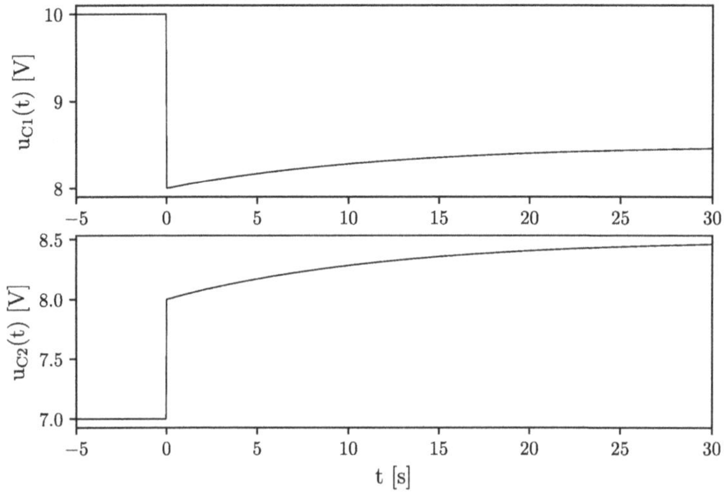

Fig. 1.112 Time evolution of $u_{C1}(t)$ and $u_{C2}(t)$

1.7.22 Inductors in Series. Impulse Response

The circuit represented in Fig. 1.113 is in steady state when at $t = 0$ the switches change their position. Given that $L_1 = 4\,\mathrm{H}$ and $L_2 = 2\,\mathrm{H}$, calculate the current $i_{L2}(t)$ and the voltage $u(t)$ for $t > 0$.

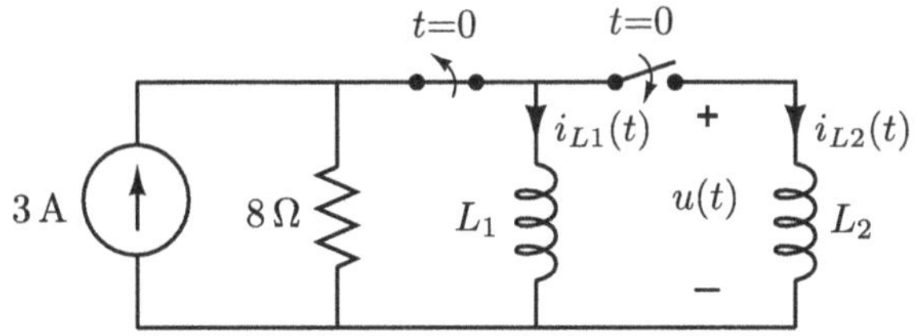

Fig. 1.113 Circuit considered in the problem under study

Solution. First, the current of each inductor will be obtained before the switches change their position, that is, at $t = 0$. It can be verified that no current flows through inductor 2 at that moment, therefore:

$$i_{L2}(0^-) = 0\,\text{A}$$

Regarding inductor 1, it will be considered that the circuit is in steady state before the switches change their position. Since, in this situation, only a dc-type excitation source is present, the dc circuit shown in Fig. 1.114 will be solved, where the inductor has been replaced by a short circuit.

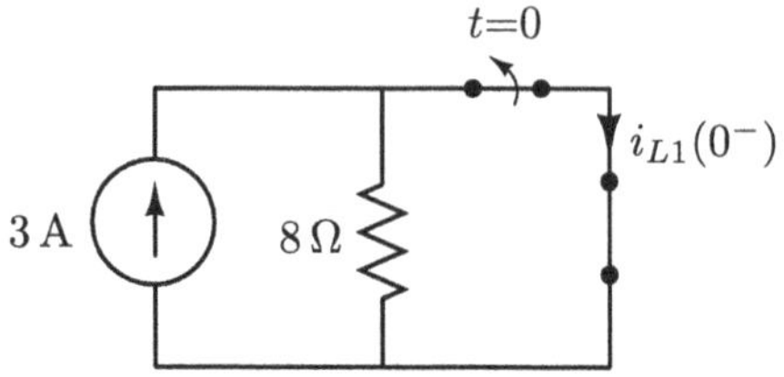

Fig. 1.114 Circuit at $t = 0^-$

According to Fig. 1.114:

$$i_{L1}(0^-) = 3\,\text{A}$$

Before the switches change their position, the current flowing through each inductor is different, specifically 3 A through inductor 1, while no current flows through inductor 2. When the switches change their position, both inductors are connected in series, as shown in Fig. 1.115.

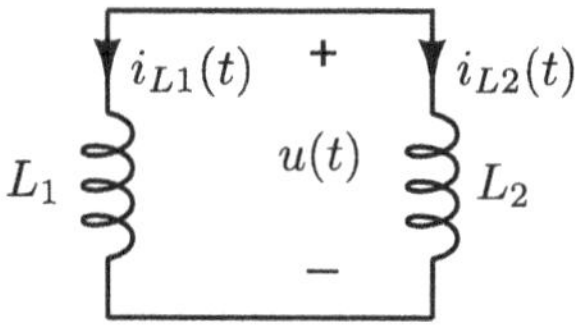

Fig. 1.115 Circuit for $t > 0$

Immediately after the position change of the switches, the same current circulates through the inductors at the expense of a voltage impulse. In this instantaneous process, the total flux in both inductors does not change. Considering $i_{L2}(t)$ as the resulting intensity reference, then:

$$i_{L2}(0^+) = \frac{-L_1 \cdot i_{L1}(0^-) + L_2 \cdot i_{L2}(0^-)}{L_1 + L_2} = \frac{-4 \cdot 3 + 2 \cdot 0}{4 + 2} = -2\,\text{A}$$

It is important to remark that, in the previous expression, it is necessary to change the sign of the current $i_{L1}(0^-)$, due to its sense is opposite to that of the current $i_{L2}(t)$, which is taken as reference.

In this particular case, for $t > 0$, the current in both inductors remains constant (Fig. 1.116):

$$i_{L1}(t) = 2\,\text{A} \ ; \ i_{L2}(t) = -2\,\text{A}$$

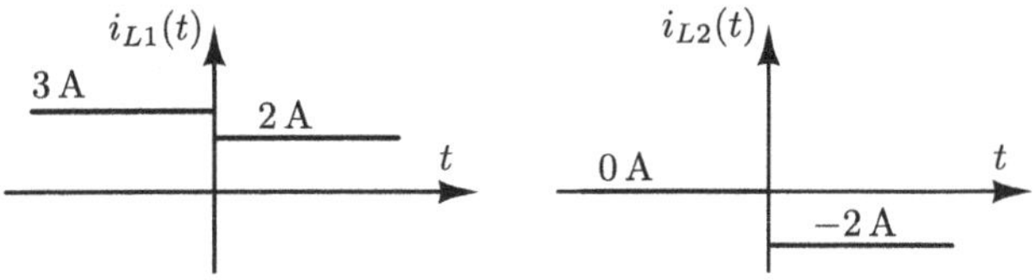

Fig. 1.116 Time evolution of $i_{L1}(t)$ and $i_{L2}(t)$

On the other hand, according to the polarity references shown in Fig. 1.115, the relation between voltage and current of the inductor 2 is:

$$u(t) = L_2 \cdot \frac{di_{L2}(t)}{dt}$$

This relation can be expressed as follows:

$$u(t)dt = L_2 \cdot di_{L2}(t)$$

Integrating the previous expression between the infinitesimal limits of 0^+ and 0^-, when the current discontinuity occurs in both inductors, the result is:

$$\int_{0^-}^{0^+} u(t)dt = \int_{0^-}^{0^+} L_2 \cdot di_{L2}(t) = L_2 \cdot \left[i_{L2}(0^+) - i_{L2}(0^-)\right] = 2 \cdot [-2 - 0] = -4\,\text{Wb}$$

Using the definition of the Dirac delta function, $u(t)$ is obtained for $t > 0$:

$$u(t) = -4 \cdot \delta(t)\,\text{V}$$

1.7.23 RL. Impulse Response Under AC Supply

The circuit represented in Fig. 1.117 is in steady state when, at $t = 0$, the switch k1 is closed and the switch k2 is opened. Given that $L_1 = 0{,}3\,\text{H}$, $L_2 = 0{,}1\,\text{H}$, $e_{g1}(t) = 50\sin(10t)\,\text{V}$ and $e_{g2}(t) = 25\sin(20t)\,\text{V}$, calculate the current $i_{L2}(t)$ for $t > 0$.

Solution. First, the current of each inductor will be obtained before the switches change their position, that is, at instant $t = 0^-$. It can be verified for inductor 1 that no current is circulating at that moment, therefore:

$$i_{L1}(0^-) = 0\,\text{A}$$

Regarding inductor 2, it will be considered that the circuit is in steady state before the switches change position. Since, in this situation, only an ac-type excitation

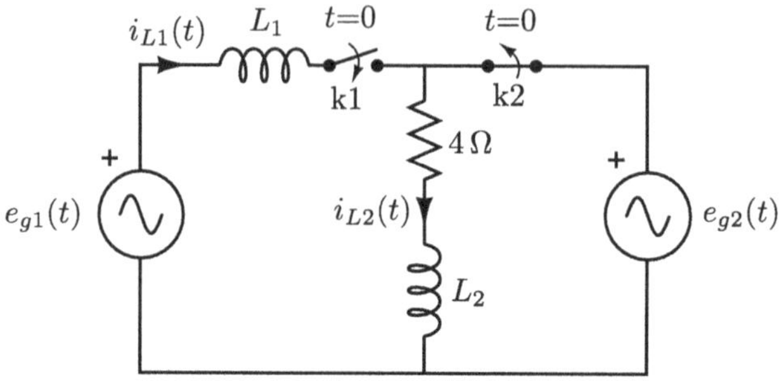

Fig. 1.117 Circuit considered in the problem under study

source is present, the circuit in the phasor domain shown in Fig. 1.118 will be solved, from which the following result is obtained:

$$\overline{I}_{L2} = \frac{\frac{25}{\sqrt{2}}\angle 0^\circ}{4+2j} \approx 3{,}95\angle -26{,}57^\circ \text{ A}$$

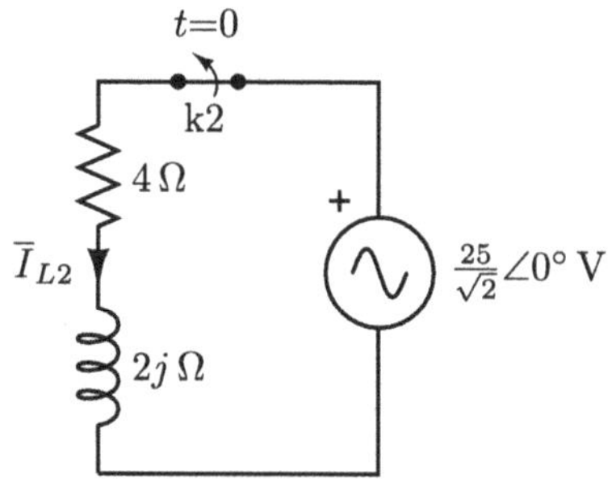

Fig. 1.118 Circuit in the phasor domain

Once the current in the phasor domain has been calculated, the corresponding expression in the time domain is derived

$$i_{L2}(t) = 3{,}95\sqrt{2}\sin(20t - 26{,}57^\circ) \text{ A}$$

whose value for the instant $t = 0^-$ is the following:

$$i_{L2}(0^-) = 3{,}95\sqrt{2}\sin(-26{,}57^\circ) \approx -2{,}5 \text{ A}$$

Before switch k1 closes and switch k2 opens, the current flowing through each inductor is different, specifically the current flowing through inductor 1 is zero while that of inductor 2 is −2.5 A. When the switches change their position, both inductors are connected in series, as shown in Fig. 1.119.

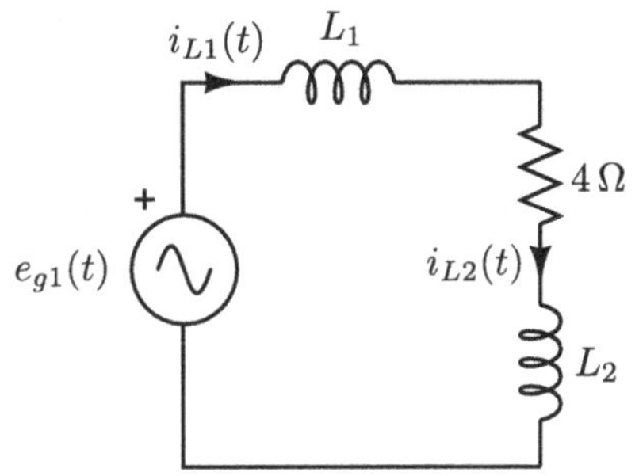

Fig. 1.119 Circuit for $t > 0$

Immediately after the position change of the switches, the same current will circulate through the inductors, at the expense of a voltage impulse. In this instantaneous process, the total flux in both inductors does not change. Taking $i_{L2}(t)$ as the reference of the resulting current:

$$i_{L2}(0^+) = \frac{L_1 \cdot i_{L1}(0^-) + L_2 \cdot i_{L2}(0^-)}{L_1 + L_2} = \frac{0{,}3 \cdot 0 - 0{,}1 \cdot 2{,}5}{0{,}3 + 0{,}1} = -0{,}625\,\text{A}$$

From $t = 0^+$, the inductors are connected in series with an initial current of -0.625 A. In this situation, both inductors can be replaced by a single equivalent inductor with a self-inductance coefficient:

$$L_{eq} = L_1 + L_2 = 0{,}3 + 0{,}1 = 0{,}4\,\text{H}$$

and an initial current of -0.625 A. This way, the resulting circuit is that shown in Fig. 1.120, where

$$i_{L2}(0^+) = -0{,}625\,\text{A}$$

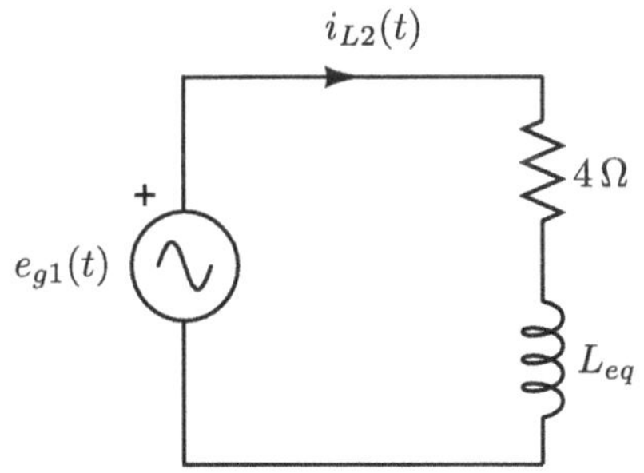

Fig. 1.120 Circuit for $t > 0$

The steady-state value of the current $i_{L2}(t)$ is subsequently obtained. In this case, an ac-type excitation source is present, so that the circuit in the phasor domain shown in Fig. 1.121 will be solved.

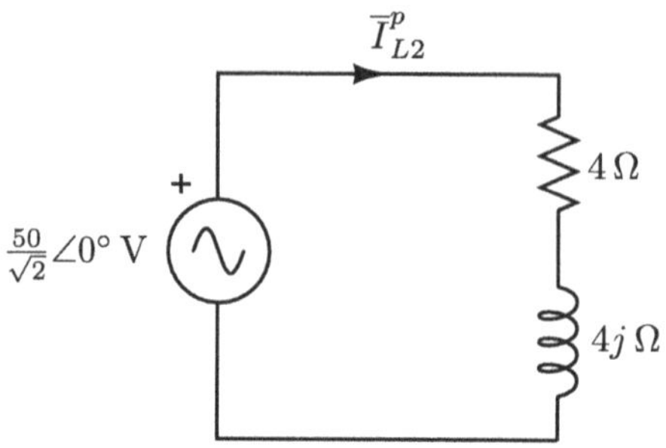

Fig. 1.121 Circuit in phasor domain

From the circuit in Fig. 1.121:

$$\overline{I}^p_{L2} = \frac{\frac{50}{\sqrt{2}}\angle 0°}{4+4j} = 6{,}25\angle -45°\ \text{A}$$

Once the steady-state current is obtained in the phasor domain, the corresponding expression in the time domain will be obtained:

$$i^p_{L2}(t) = 6{,}25\sqrt{2}\sin(10t - 45°)\ \text{A}$$

whose value for $t = 0^+$ is:

$$i^p_{L2}(0^+) = 6{,}25\sqrt{2}\sin(-45°) = -6{,}25\ \text{A}$$

Being an RL circuit, the time constant (for $t > 0$) is calculated as follows:

$$\tau = \frac{L_{eq}}{R} = \frac{0{,}4}{4} = \frac{1}{10}\ \text{s}$$

Finally, once the initial condition, the steady-state response and the time constant have been obtained, the complete response of $i_{L2}(t)$ is obtained for $t > 0$:

$$\begin{aligned} i_{L2}(t) &= i^p_{L2}(t) + \left[i_{L2}(0^+) - i^p_{L2}(0^+)\right]\cdot \text{e}^{-t/\tau} \\ &= 6{,}25\sqrt{2}\sin(10t - 45°) + [-0{,}625 + 6{,}25]\cdot \text{e}^{-10t}\ \text{A} \end{aligned}$$

The time evolution of $i_{L2}(t)$ has been graphically represented in Fig. 1.122.

1.7.24 RC. Thévenin Equivalent Under DC Supply

The switch in the circuit represented in Fig. 1.123 closes at $t = 0$. Given that $u_C(0^-) = 6$ V, calculate the voltage $u_C(t)$ for $t > 0$.

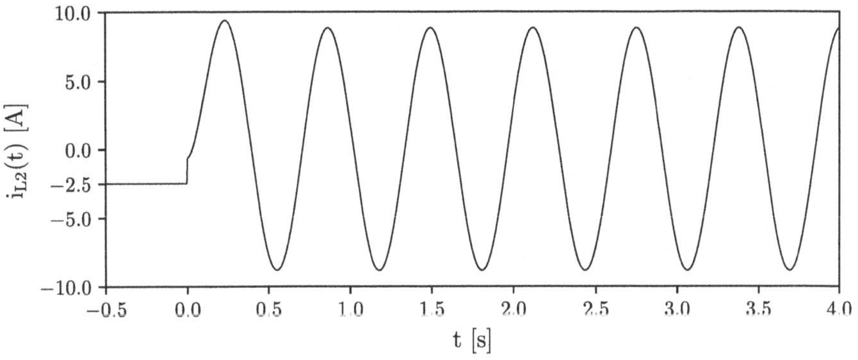

Fig. 1.122 Time evolution of $i_{L2}(t)$

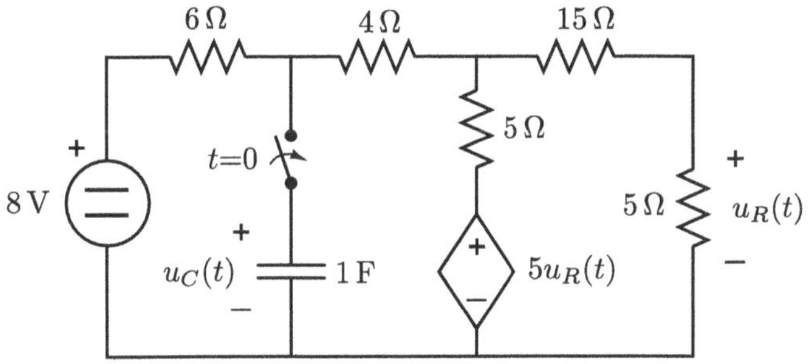

Fig. 1.123 Circuit considered in the problem under study

Solution. To simplify the study of the transient, the Thévenin equivalent seen from the capacitor terminals (a and b) will be obtained, as shown in Fig. 1.124.

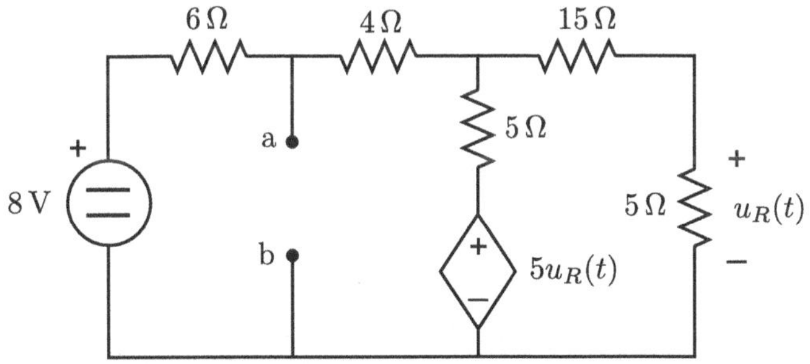

Fig. 1.124 Circuit seen from the capacitor terminals

To obtain the Thévenin equivalent, the open-circuit voltage will be first calculated, then the short circuit current will be obtained. Regarding the calculation of the open-circuit voltage between terminals a and b, the circuit to be solved is that shown in Fig. 1.125. Mesh analysis method will be used for the resolution.

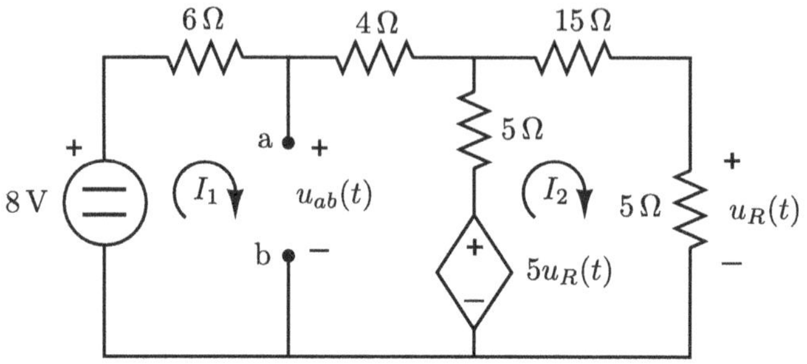

Fig. 1.125 Circuit to obtain the open-circuit voltage between terminals **a** and **b**

From the circuit in Fig. 1.125:

$$\begin{bmatrix} 15 & -5 \\ -5 & 25 \end{bmatrix} \begin{bmatrix} I_1 \\ I_2 \end{bmatrix} = \begin{bmatrix} 8 - 5u_R(t) \\ 5u_R(t) \end{bmatrix}$$

Next, it is necessary to establish the relation between the control variable of the dependent source, $u_R(t)$, and the mesh currents. According to the circuit in Figure Fig. 1.125:

$$u_R(t) = 5 \cdot I_2$$

Thus:

$$\begin{bmatrix} 15 & -5 \\ -5 & 25 \end{bmatrix} \begin{bmatrix} I_1 \\ I_2 \end{bmatrix} = \begin{bmatrix} 8 - 5 \cdot 5 \cdot I_2 \\ 5 \cdot 5 \cdot I_2 \end{bmatrix}$$

Rearranging terms, the mesh equations are finally obtained:

$$\begin{bmatrix} 15 & 20 \\ -5 & 0 \end{bmatrix} \begin{bmatrix} I_1 \\ I_2 \end{bmatrix} = \begin{bmatrix} 8 \\ 0 \end{bmatrix}$$

The mesh currents are obtained solving the previous equation:

$$I_1 = 0\,\text{A} \; ; \; I_2 = \frac{8}{20}\,\text{A}$$

Therefore, the voltage $u_{ab}(t)$ is calculated as follows:

$$u_{ab}(t) = 8 - 6 \cdot I_1 = 8\,\text{V}$$

On the other hand, the short-circuit current between terminals a and b will then be calculated, as shown in the circuit represented in Fig. 1.126.

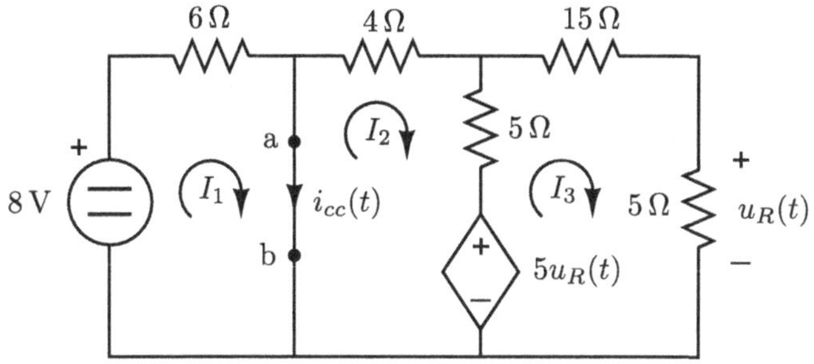

Fig. 1.126 Circuit to obtain short-circuit current between terminals **a** and **b**

Mesh analysis method is applied again:

$$\begin{bmatrix} 6 & 0 & 0 \\ 0 & 9 & -5 \\ 0 & -5 & 25 \end{bmatrix} \begin{bmatrix} I_1 \\ I_2 \\ I_3 \end{bmatrix} = \begin{bmatrix} 8 \\ -5u_R(t) \\ 5u_R(t) \end{bmatrix}$$

A relation is established between the control variable of the dependent source, $u_R(t)$, and the mesh currents, using the circuit in Figure Fig. 1.126:

$$u_R(t) = 5 \cdot I_3$$

Thus

$$\begin{bmatrix} 6 & 0 & 0 \\ 0 & 9 & -5 \\ 0 & -5 & 25 \end{bmatrix} \begin{bmatrix} I_1 \\ I_2 \\ I_3 \end{bmatrix} = \begin{bmatrix} 8 \\ -5 \cdot 5 \cdot I_3 \\ 5 \cdot 5 \cdot I_3 \end{bmatrix}$$

Rearranging terms, the mesh equations are finally obtained:

$$\begin{bmatrix} 6 & 0 & 0 \\ 0 & 9 & 20 \\ 0 & -5 & 0 \end{bmatrix} \begin{bmatrix} I_1 \\ I_2 \\ I_3 \end{bmatrix} = \begin{bmatrix} 8 \\ 0 \\ 0 \end{bmatrix}$$

The mesh currents are consequently calculated:

$$I_1 = \frac{8}{6}\,\text{A} \; ; \; I_2 = 0\,\text{A} \; ; \; I_3 = 0\,\text{A}$$

Therefore, the current $i_{cc}(t)$ is:

$$i_{cc}(t) = I_1 - I_2 = \frac{8}{6}\,\text{A}$$

Once the open-circuit voltage and the short-circuit current between terminals a and b have been obtained, the equivalent resistance (Thévenin resistance) is calculated as follows:

$$R_{eq} = \frac{u_{ab}(t)}{i_{cc}(t)} = \frac{8}{8/6} = 6\,\Omega$$

With the previous results, the RC circuit in Fig. 1.127 is reached.

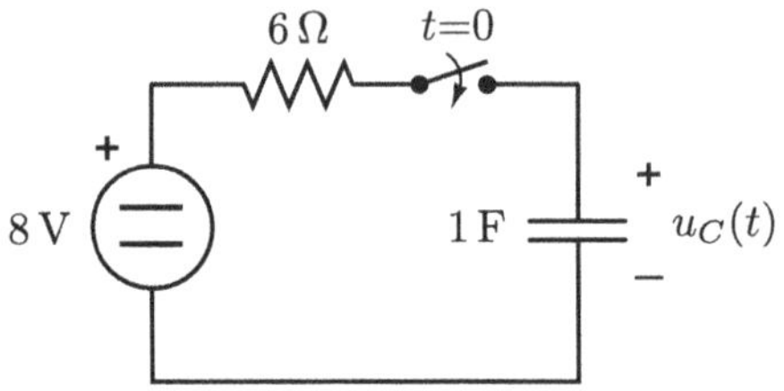

Fig. 1.127 Resultant RC circuit

According to the problem statement, before closing the switch ($t = 0^-$), the capacitor voltage is $u_C(0^-) = 6\,\text{V}$. Except for an impulse-type response, the capacitor voltage does not change when the switch is closed, therefore:

$$u_C(0^+) = u_C(0^-) = 6\,\text{V}$$

The steady-state capacitor voltage, $u_C^p(t)$ will be subsequently obtained. In this case, once the switch is closed, only a dc-type source of excitation is present, so that the dc circuit shown in Fig. 1.128 will be solved, where the capacitor has been replaced by a open circuit.

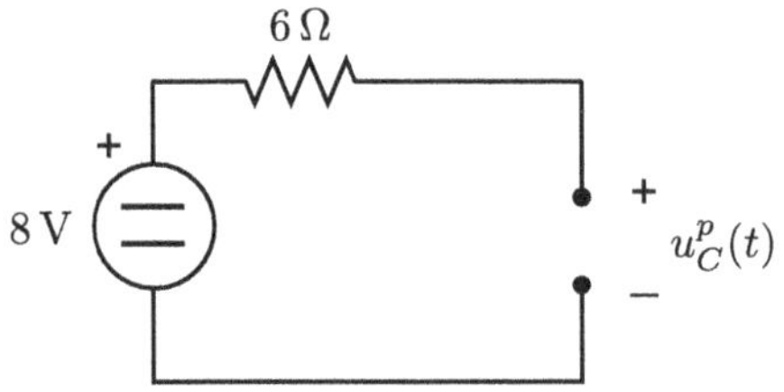

Fig. 1.128 Circuit in dc steady-state regime

From the circuit in Fig. 1.128:

$$u_C^p(t) = 8\,\text{V}$$

Substituting $t = 0^+$:

$$u_C^p(0^+) = 8\,\text{V}$$

Being an RC type circuit, the time constant is calculated as:

$$\tau = R_{eq}C = 6 \cdot 1 = 6\,\text{s}$$

Finally, the complete reponse of $u_C(t)$ for $t > 0$ is consequently derived:

$$u_C(t) = u_C^p(t) + \left[u_C(0^+) - u_C^p(0^+)\right] \cdot e^{-t/\tau} = 8 + [6 - 8] \cdot e^{-t/6} = 8 - 2 \cdot e^{-t/6}\ \mathrm{V}$$

1.7.25 RC. Thévenin Equivalent Under DC Supply

The circuit represented in Fig. 1.129 is in steady state when the switch is closed at $t = 0$. Calculate the voltage $u_C(t)$ for $t > 0$.

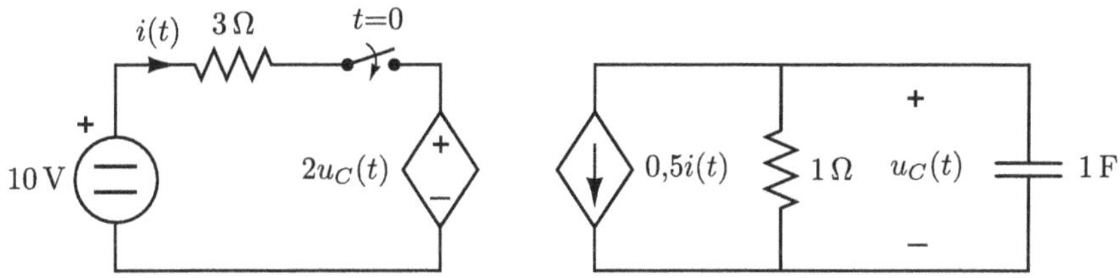

Fig. 1.129 Circuit considered in the problem under study

Solution. To simplify the study of the transient, the Thévenin equivalent seen from the terminals of the capacitor (with the switch closed) will be obtained, as shown in Fig. 1.130. For the sake of clarity, the control variable $u_C(t)$ has been replaced by $u_\alpha(t)$.

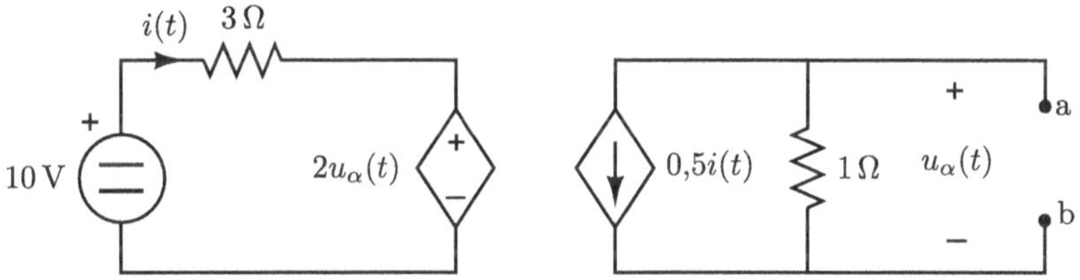

Fig. 1.130 Circuit seen from the terminals of the capacitor

To obtain the Thévenin equivalent, the open-circuit voltage will be first calculated and subsequently the short circuit current. Regarding the calculation of the open-circuit voltage between terminals a and b, the circuit to be solved is that shown in Fig. 1.131.

From the circuit in Fig. 1.131:

$$\left.\begin{array}{r} 10 = 3i(t) + 2u_\alpha(t) \\ u_\alpha(t) = -0{,}5i(t) \end{array}\right\} \Rightarrow u_\alpha(t) = \frac{-5}{2}\ \mathrm{V}$$

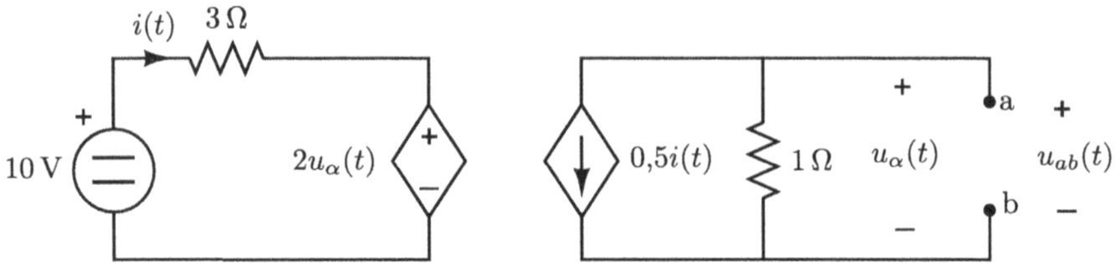

Fig. 1.131 Circuit to obtain the open-circuit voltage between terminals a and b

Therefore, the open-circuit voltage between terminals a and b is:

$$u_{ab}(t) = u_\alpha(t) = -2{,}5\,\mathrm{V}$$

On the other hand, the short-circuit current between terminals a and b will be calculated as shown in the circuit in Fig. 1.132.

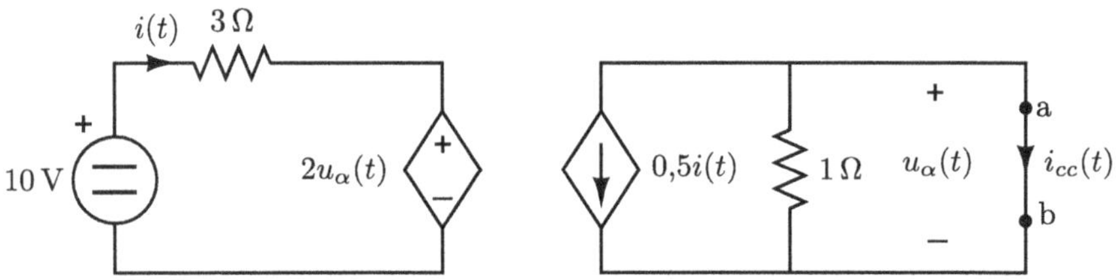

Fig. 1.132 Circuit to obtain the short-circuit current between terminals a and b

From the circuit in Fig. 1.132:

$$\left.\begin{aligned} 10 &= 3i(t) + 2u_\alpha(t) \\ u_\alpha(t) &= 0 \\ i_{cc}(t) &= -0{,}5i(t) \end{aligned}\right\} \Rightarrow \ i_{cc}(t) = \frac{-5}{3}\,\mathrm{A}$$

Once the open-circuit voltage and the short-circuit current between terminals a and b have been obtained, the equivalent resistance (Thévenin resistance) is calculated as follows:

$$R_{eq} = \frac{u_{ab}(t)}{i_{cc}(t)} = \frac{-5/2}{-5/3} = 1{,}5\,\Omega$$

This way, for $t > 0$, the RC circuit represented in Fig. 1.133 is reached.

According to the problem statement, before the switch closes ($t = 0^-$), the circuit is in steady state. In this situation it can be verified that $u_C(0^-) = 0\,\mathrm{V}$. Except for an impulse-type response, the capacitor voltage- does not change when the switch is closed, therefore:

$$u_C(0^+) = u_C(0^-) = 0\,\mathrm{V}$$

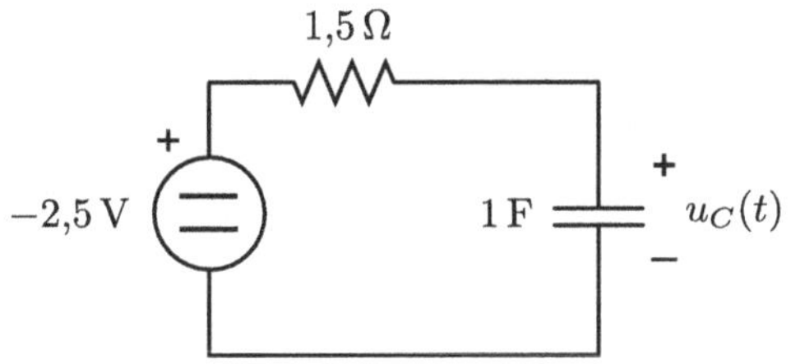

Fig. 1.133 Resultant RC circuit

The steady-state capacitor voltage, $u_C^p(t)$ will be subsequently obtained. In this case, once the switch is closed, only a dc-type source of excitation is present, so that the dc circuit shown in Fig. 1.134 will be solved, where the capacitor has been replaced by a open circuit.

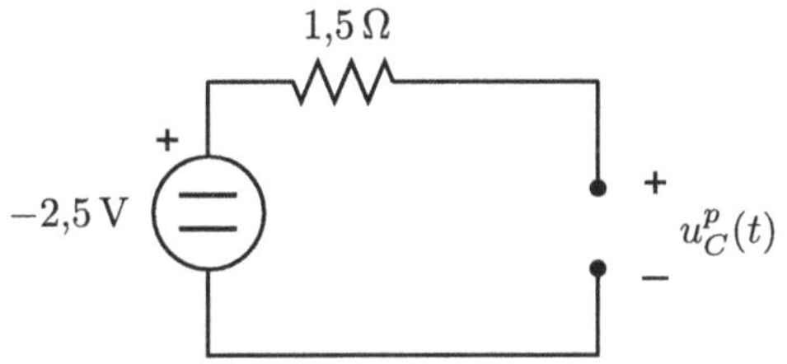

Fig. 1.134 Circuit in dc steady-state regime

From the circuit in Fig. 1.134:

$$u_C^p(t) = -2{,}5\,\text{V}$$

Substituting $t = 0^+$:

$$u_C^p(0^+) = -2{,}5\,\text{V}$$

Being an RC circuit, the time constant is calculated as:

$$\tau = RC = 1{,}5 \cdot 1 = \frac{3}{2}\,\text{s}$$

Finally, the complete response of $u_C(t)$ for $t > 0$ is consequently derived:

$$\begin{aligned} u_C(t) &= u_C^p(t) + \left[u_C(0^+) - u_C^p(0^+)\right] \cdot \text{e}^{-t/\tau} \\ &= -2{,}5 + [0 + 2{,}5] \cdot \text{e}^{-2t/3} = -2{,}5 \cdot (1 - \text{e}^{-2t/3})\,\text{V} \end{aligned}$$

1.7.26 RC. Infinite Time Constant Under AC Supply

The circuit represented in Fig. 1.135 is in steady state when, at $t = 0$, the switches change their position. Given that $i_g(t) = 6\cos(10t + 30°)$ A, calculate the voltage $u_C(t)$ for $t > 0$.

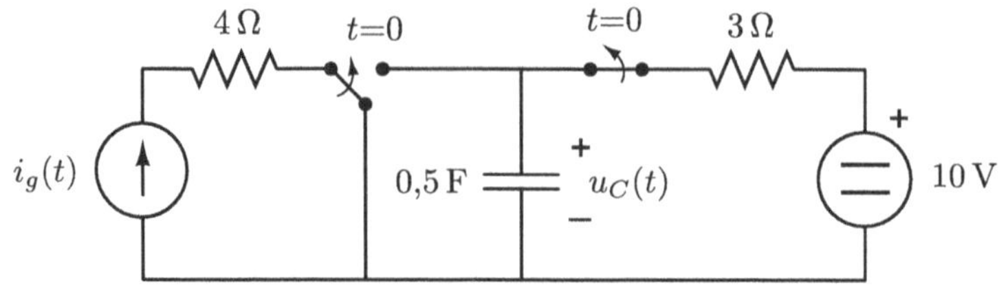

Fig. 1.135 Circuit considered in the problem under study

Solution. First, the capacitor voltage will be obtained before the switches change their position ($t = 0^-$). For this purpose, it will be considered that the circuit is in steady-state regime before this change. Since, in this situation, only a dc-type source of excitation is present, the dc circuit shown in Fig. 1.136 will be solved, where the capacitor has been replaced by an open circuit.

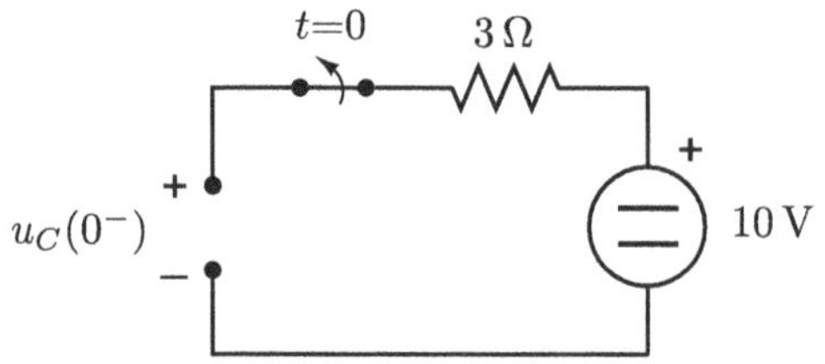

Fig. 1.136 Circuit at $t = 0^-$

From Fig. 1.136 the capacitor voltage at $t = 0^-$ can be obtained:

$$u_C(0^-) = 10\,\text{V}$$

Except for an impulse-type response, which is not the case, the capacitor voltage does not shift when the switches change their position, therefore:

$$u_C(0^+) = u_C(0^-) = 10\,\text{V}$$

Once the switches change their position, the resulting circuit is that shown in Fig. 1.137. The reader may notice that the current source connected in series with the resistance has been replaced by an equivalent, remaining exclusively the electrical source.

The voltage $u_C(t)$ can be obtained as follows:

$$u_C(t) = u_C^p(t) + \left[u_C(0^+) - u_C^p(0^+)\right] \cdot \text{e}^{-t/\tau}$$

In this particular case, as shown in Fig. 1.137, the circuit has an infinite resistance seen from the capacitor terminals, so that the time constant is also infinite. It should

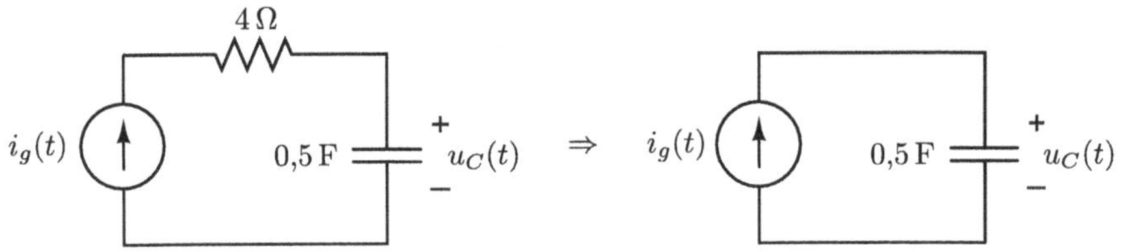

Fig. 1.137 Circuit for $t > 0$

be noted that, to calculate the equivalent resistance, a passive circuit must be derived, so that the capacitor would remain open circuit. Being this time constant, then:

$$u_C(t) = u_C^p(t) + u_C(0^+) - u_C^p(0^+)$$

The steady-state capacitor voltage, $u_C^p(t)$ will then subsequently obtained. In this case, since only an ac-type source of excitation is present, the circuit will be solved in the phasor domain, as shown in Fig. 1.138.

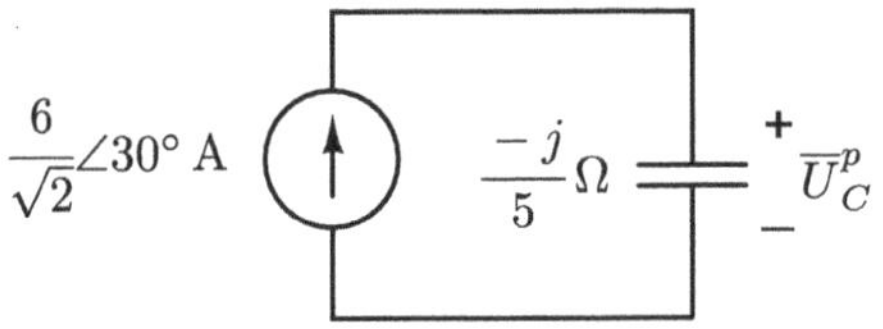

Fig. 1.138 Circuit in the phasor domain

From the circuit in Fig. 1.138:

$$\overline{U}_C^p = \frac{6}{\sqrt{2}}\angle 30° \cdot \frac{-j}{5} = \frac{6}{5 \cdot \sqrt{2}}\angle -60° \text{ V}$$

Once the steady-state voltage in the phasor domain has been obtained, the corresponding expression in the time domain will be obtained:

$$u_C^p(t) = \frac{6}{5} \cdot \cos(10t - 60°) \text{ V}$$

whose value for $t = 0^+$ is:

$$u_C^p(0^+) = \frac{6}{5} \cdot \cos(-60°) = \frac{6}{10} \text{ V}$$

So:

$$u_C(t) = u_C^p(t) + u_C(0^+) - u_C^p(0^+) = \frac{6}{5} \cdot \cos(10t - 60°) + 10 - \frac{6}{10}$$
$$= 1{,}2 \cdot \cos(10t - 60°) + 9{,}4$$

An alternative way to solve the problem is the analysis of the circuit in Fig. 1.137, according to which [3]:

$$u_C(t) = u_C(0^+) + \frac{1}{0{,}5}\int_{0^+}^{t} i_g(\lambda)d\lambda$$

Substituting the known values:

$$u_C(t) = 10 + \frac{1}{0{,}5}\int_{0^+}^{t} 6\cos(10\lambda + 30°)d\lambda = 10 + 12 \cdot \frac{1}{10} \cdot \left[\sin(10t + 30°) - \sin 30°\right]$$

Resulting:

$$u_C(t) = 10 + 1{,}2 \cdot \left[\sin(10t + 30°) - \sin 30°\right] = 9{,}4 + 1{,}2 \cdot \sin(10t + 30°)\text{ V}$$

In Fig. 1.139, the time evolution of $u_C(t)$ has been graphically represented.

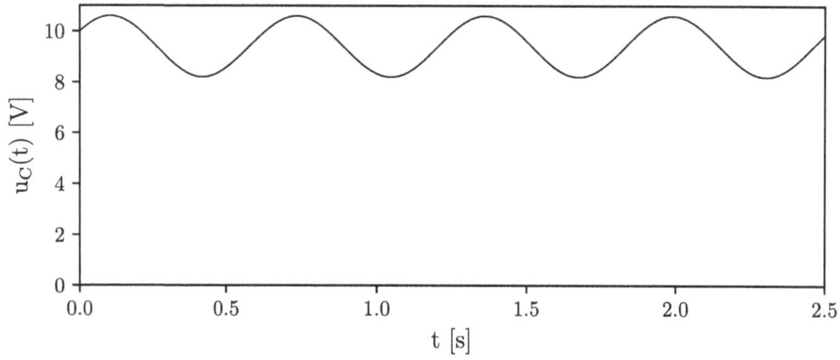

Fig. 1.139 Time evolution of $u_C(t)$

1.7.27 RC. Impulse Response, Concatenated Transients, Infinite Time Constant, DC Supply

In the circuit represented in Fig. 1.140 it is known that $u_{C1}(0^-) = 0\,\text{V}$, $u_{C2}(0^-) = 5\,\text{V}$, $C_1 = 3\,\text{F}$ and $C_2 = 2\,\text{F}$. Calculate the voltage $u_{C1}(t)$ for $t > 0$.

Solution. According to the circuit in Fig. 1.140, the first switch closes at $t = 0$, the second at $t = 1\,\text{s}$ and the third one opens at $t = 2\,\text{s}$. Therefore, a first circuit in the time interval $0 < t < 1\,\text{s}$ will be analyzed, a second circuit in the interval $1\,\text{s} < t < 2\,\text{s}$, and a third circuit in the interval $t > 2\,\text{s}$.

[3] The variable λ has been used to prevent the integration variable from matching the upper limit.

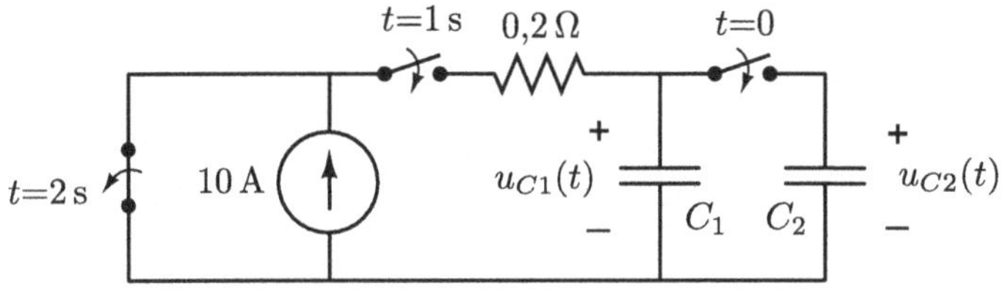

Fig. 1.140 Circuit considered in the problem under study

Circuit analysis for $0 < t < 1\,\mathrm{s}$

According to the problem statement, before the first switch closes at $t = 0$, one of the capacitors is charged while the other is discharged. When the switch is closed, both capacitors are connected in parallel. Immediately after this closing, the capacitors are subjected to the same voltage at the expense of an intensity pulse. In this instantaneous process, the total charge stored in both capacitors does not change, which allows the calculation of the common voltage as:

$$u_{C1}(0^+) = u_{C2}(0^+) = \frac{C_1 \cdot u_{C1}(0^-) + C_2 \cdot u_{C2}(0^-)}{C_1 + C_2} = \frac{3 \cdot 0 + 2 \cdot 5}{3 + 2} = \frac{10}{5} = 2\,\mathrm{V}$$

For $0 < t < 1\,\mathrm{s}$, both capacitor voltages remain constant since no current flows through these elements:

$$u_{C1}(t) = u_{C2}(t) = 2\,\mathrm{V}$$

From $t = 0^+$, the capacitors are connected in parallel with an initial voltage of 2 V. In this situation, both capacitors can be replaced by a single one whose capacity is

$$C_{eq} = C_1 + C_2 = 3 + 2 = 5\,\mathrm{F}$$

while the corresponding initial voltage is 2 V. This way, the circuit for $0 < t < 1\,\mathrm{s}$ is that shown in Fig. 1.141.

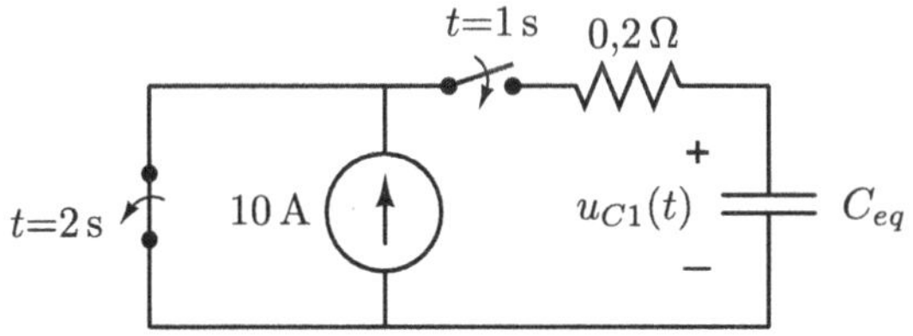

Fig. 1.141 Circuit for $0 < t < 1\,\mathrm{s}$

Before the second switch closes at $t = 1\,\mathrm{s}$, the voltage across the equivalent capacitor is:

$$u_{C1}(1\,\mathrm{s}^-) = 2\,\mathrm{V}$$

Circuit analysis for $1\,\mathrm{s} < t < 2\,\mathrm{s}$
When the second switch is closed at $t = 1\,\mathrm{s}$, the resulting circuit is that shown in Fig. 1.142.

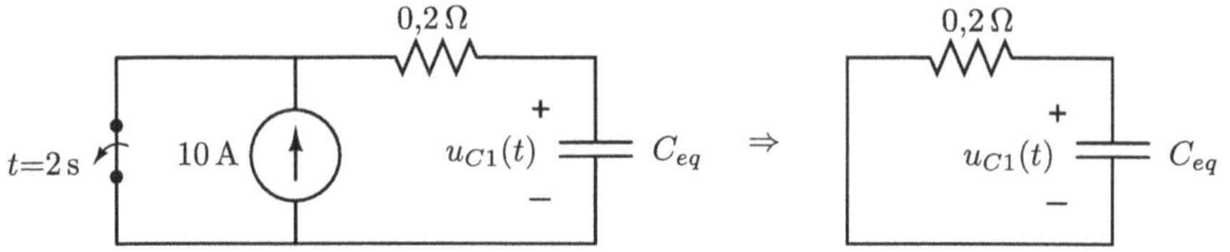

Fig. 1.142 Circuit for $1\,\mathrm{s} < t < 2\,\mathrm{s}$

Except for an impulse-type response, which is not the case, the capacitor voltage immediately after the second switch closes has the same value as that immediately before, therefore:

$$u_{C1}(1\,\mathrm{s}^+) = u_{C1}(1\,\mathrm{s}^-) = 2\,\mathrm{V}$$

The steady-state capacitor voltage, $u_C^p(t)$ will be subsequently obtained. In this case there is no excitation source in the resulting circuit, only the capacitor connected to the resistance. Consequently:

$$u_C^p(t) = 0\,\mathrm{V} \;\; ; \;\; u_C^p(1\,\mathrm{s}^+) = 0\,\mathrm{V}$$

Being an RC-type circuit, the time constant is calculated as:

$$\tau = RC_{eq} = 0{,}2 \cdot 5 = 1\,\mathrm{s}$$

The response of the voltage $u_{C1}(t)$ for $1 < t < 2\,\mathrm{s}$ is consequently obtained:

$$\begin{aligned} u_{C1}(t) &= u_{C1}^p(t) + \left[u_{C1}(1\,\mathrm{s}^+) - u_{C1}^p(1\,\mathrm{s}^+)\right] \cdot \mathrm{e}^{-(t-1)/\tau} \\ &= 0 + [2 - 0] \cdot \mathrm{e}^{-(t-1)/1} = 2 \cdot \mathrm{e}^{-(t-1)/1}\,\mathrm{V} \end{aligned}$$

Before the third switch opens at $t = 2\,\mathrm{s}$, the voltage across the equivalent capacitor is:

$$u_{C1}(2\,\mathrm{s}^-) = 2 \cdot \mathrm{e}^{-(2-1)/1} \approx 0{,}735\,\mathrm{V}$$

Circuit analysis for $t > 2\,\mathrm{s}$
When the third switch is opened at $t = 2\,\mathrm{s}$, the resulting circuit is that shown in Fig. 1.143, where it can be seen that the current source connected in series with the resistance has been replaced by its equivalent.

From the circuit in Fig. 1.143:

$$u_{C1}(t) = u_{C1}(2\,\mathrm{s}^+) + \frac{1}{C_{eq}} \int_{2\,\mathrm{s}^+}^{t} 10 \cdot d\lambda$$

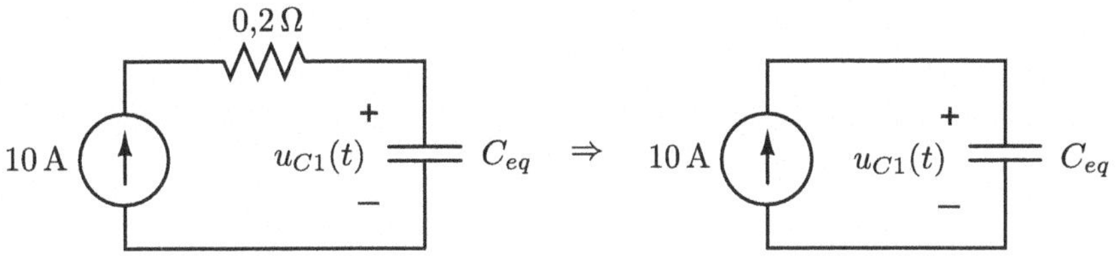

Fig. 1.143 Circuit for $t > 2\,\text{s}$

Given that

$$u_{C1}(2\,\text{s}^+) = u_{C1}(2\,\text{s}^-) = 0{,}735\,\text{V}$$

the response of the voltage $u_{C1}(t)$ for $t > 2\,\text{s}$ yields:

$$u_{C1}(t) = 0{,}735 + \frac{1}{5}\int_{2\,\text{s}^+}^{t} 10 \cdot d\lambda = 0{,}735 + \frac{10}{5} \cdot (t-2) = 0{,}735 + 2 \cdot (t-2)\,\text{V}$$

The time evolution of $u_{C1}(t)$ has been graphically represented in Fig. 1.144.

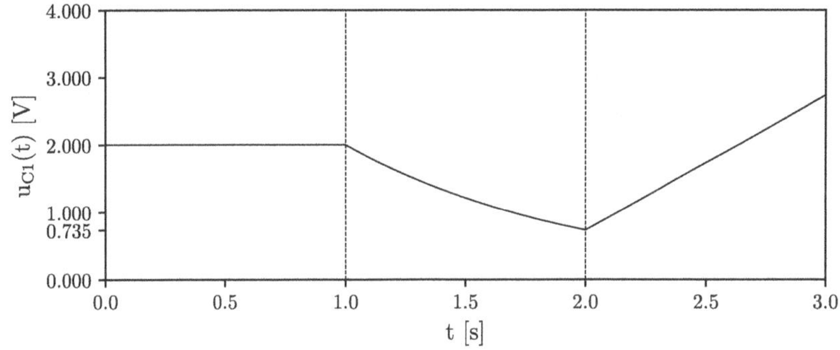

Fig. 1.144 Time evolution of $u_{C1}(t)$

1.7.28 RL. Infinite Time Constant Under AC Supply

The circuit represented in Fig. 1.145 is in steady state when, at $t = 0$, the switches change their position. Given that $u_g(t) = 100\cos(10t)\,\text{V}$, calculate the current $i_L(t)$ for $t > 0$.

Solution. First, the inductor current will be obtained before the switches change their position ($t = 0^-$). For this purpose, it will be considered that the circuit is in steady-state regime. Since, in this situation, only a dc-type excitation source is present, the dc circuit shown in Fig. 1.146 will be solved, where the inductor has been replaced by a short circuit.

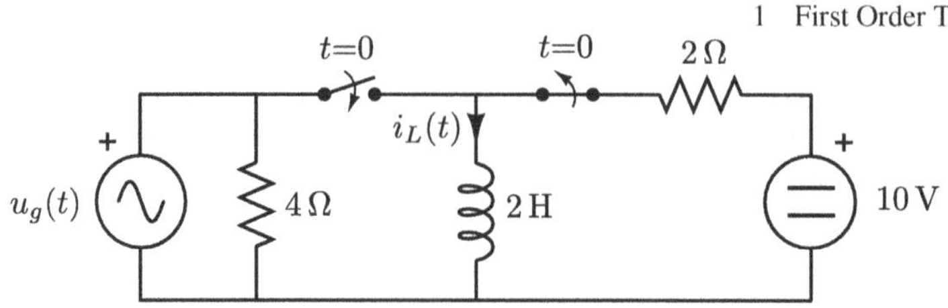

Fig. 1.145 Circuit considered in the problem under study

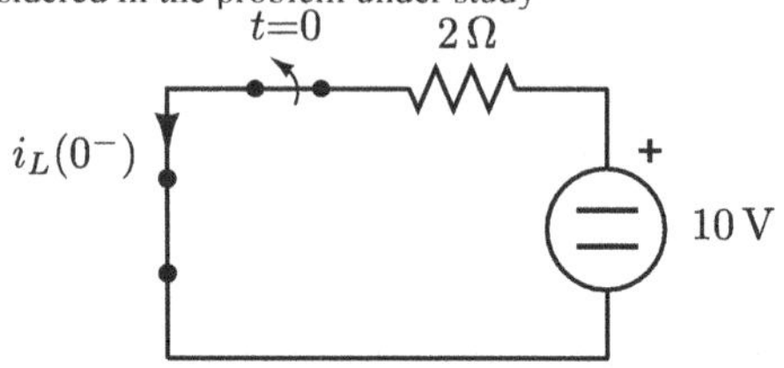

Fig. 1.146 Circuit at $t = 0^-$

From the circuit in Fig. 1.146 the current through the inductor is obtained at $t = 0^-$:

$$i_L(0^-) = \frac{10}{2} = 5\,\text{A}$$

Except for an impulse-type response, which is not the case, the inductor current does not change when the switches shift their position, therefore:

$$i_L(0^+) = i_L(0^-) = 5\,\text{A}$$

For $t > 0$, the resulting circuit is that shown in Fig. 1.147. It can be seen that the voltage source connected in parallel with the resistance has been replaced by its equivalent.

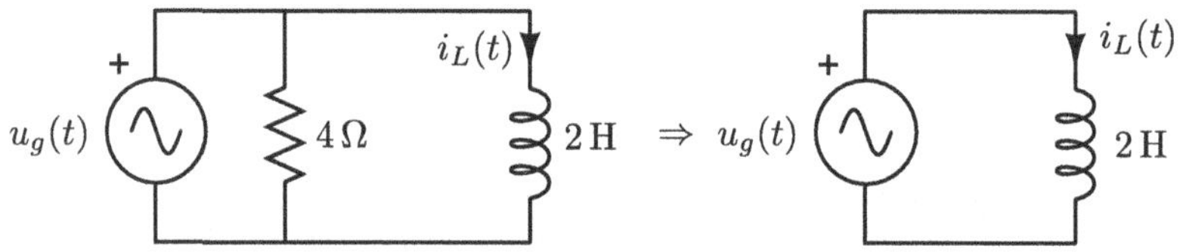

Fig. 1.147 Circuit for $t > 0$

The response of the current $i_L(t)$ can be obtained as follows:

$$i_L(t) = i_L^p(t) + \left[i_L(0^+) - i_L^p(0^+)\right] \cdot \mathrm{e}^{-t/\tau}$$

In this particular case, as shown in Fig. 1.147, the circuit has zero resistance and therefore the time constant is infinite. It can be seen that, if a passive circuit is derived, the inductor is short-circuited and therefore the equivalent resistance is zero, meaning an infinite time constant, so that:

$$i_L(t) = i_L^p(t) + i_L(0^+) - i_L^p(0^+)$$

The steady-state inductor current, $i_L^p(t)$, will be subsequently obtained. In this case, only an ac-type excitation source is present, so that the circuit shown in Fig. 1.148 will be solved in the phasor domain.

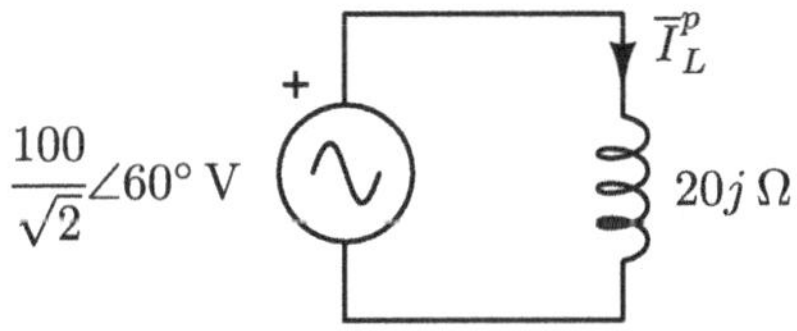

Fig. 1.148 Circuit in the phasor domain

From the circuit in Fig. 1.148:

$$\overline{I}_L^p = \frac{100}{\sqrt{2}}\angle 60° \cdot \frac{1}{20j} = \frac{5}{\sqrt{2}}\angle -30°\ \text{A}$$

Once the steady-state current is obtained in the phasor domain, the corresponding expression in the time domain will be obtained:

$$i_L^p(t) = 5\cos(10t - 30°)\ \text{A}$$

whose value for $t = 0^+$ is the following:

$$i_L^p(0^+) = 5\cos(-30°) = \frac{5\sqrt{3}}{2}\ \text{A}$$

So:

$$i_L(t) = i_L^p(t) + i_L(0^+) - i_L^p(0^+) = 5\cos(10t - 30°) + 5 - \frac{5\sqrt{3}}{2}\ \text{A}$$

An alternative way to solve the problem is the analysis of the circuit in Fig. 1.147, according to which:

$$u_g(t) = 2 \cdot \frac{di_L(t)}{dt} \Rightarrow di_L(t) = \frac{1}{2} \cdot u_g(t)dt$$

Integrating both terms of the equation results:

$$\int_{0^+}^{t} di_L(\lambda) = i_L(t) - i_L(0^+) = i_L(t) - 5$$

$$\int_{0^+}^{t} \frac{1}{2} \cdot u_g(\lambda)d\lambda = \int_{0^+}^{t} \frac{1}{2} \cdot 100\cos(10\lambda + 60°)d\lambda = \frac{50}{10} \cdot \left[\sin(10t + 60°) - \sin 60°\right]$$

So:

$$i_L(t) = 5 + 5 \cdot \left[\sin(10t + 60°) - \sin 60°\right] = 5\sin(10t + 60°) + 5 - \frac{5\sqrt{3}}{2}\ \text{A}$$

The time evolution of $i_L(t)$ has been graphically represented in Fig. 1.149.

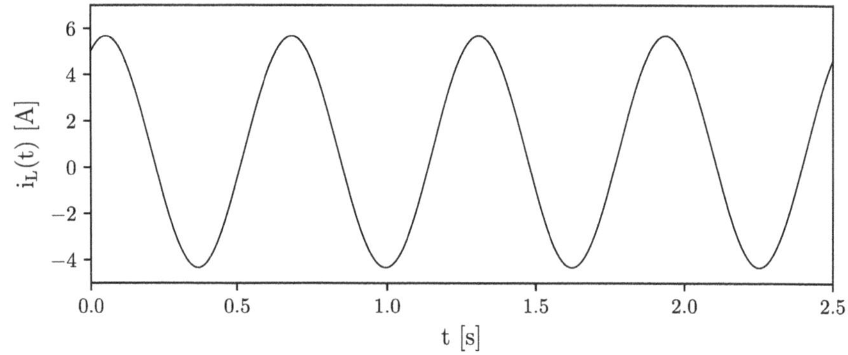

Fig. 1.149 Time evolution of $i_L(t)$

1.7.29 RL. Infinite Time Constant Under DC Supply

The circuit represented in Fig. 1.150 is in steady state when the switches change their position at $t = 0$. Calculate the current $i_L(t)$ for $t > 0$.

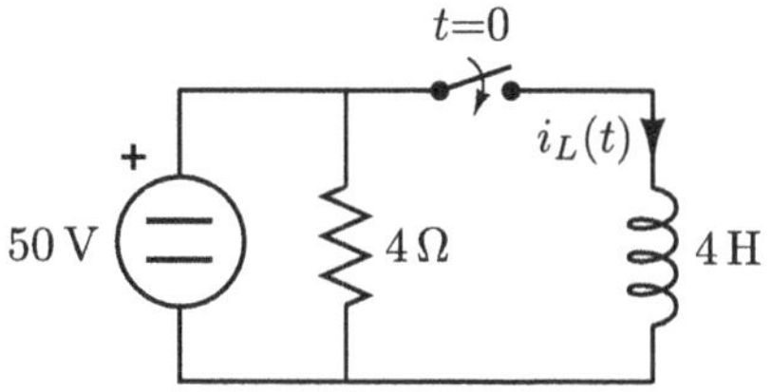

Fig. 1.150 Circuit considered in the problem under study

Solution. Before the switch closes, the current through the inductor is zero, therefore:

$$i_L(0^-) = 0\ \text{A}$$

Except for an impulse-type response, which is not the case, the current through the inductor does not change when the switch is closed, consequently:

$$i_L(0^+) = i_L(0^-) = 0\ \text{A}$$

For $t > 0$, the resulting circuit is that shown in Fig. 1.151. It can be seen that the voltage source connected in parallel with the resistance has been replaced by its equivalent, remaining exclusively the electrical source.

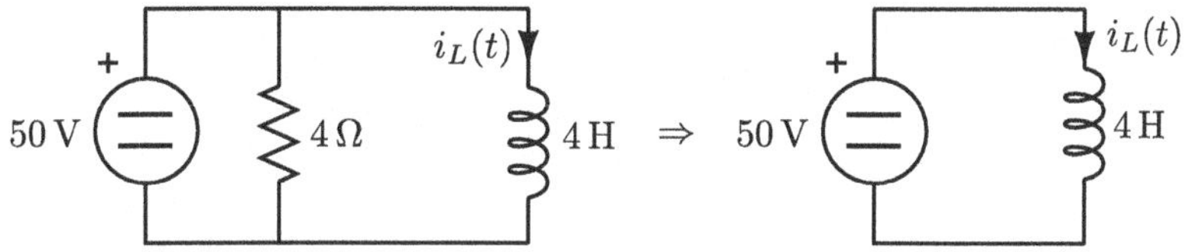

Fig. 1.151 Circuit for $t > 0$

From the circuit in Fig. 1.151:

$$50 = 4 \cdot \frac{di_L(t)}{dt} \Rightarrow di_L(t) = \frac{1}{4} \cdot 50 \cdot dt$$

Integrating both terms of the equation results:

$$\int_{0^+}^{t} di_L(\lambda) = i_L(t) - i_L(0^+) = i_L(t) - 0 = \int_{0^+}^{t} \frac{1}{4} \cdot 50 \cdot d\lambda = \frac{50}{4} \cdot t$$

So:

$$i_L(t) = \frac{50}{4} \cdot t \text{ A}$$

The time evolution of $i_L(t)$ has been graphically represented in Fig. 1.152.

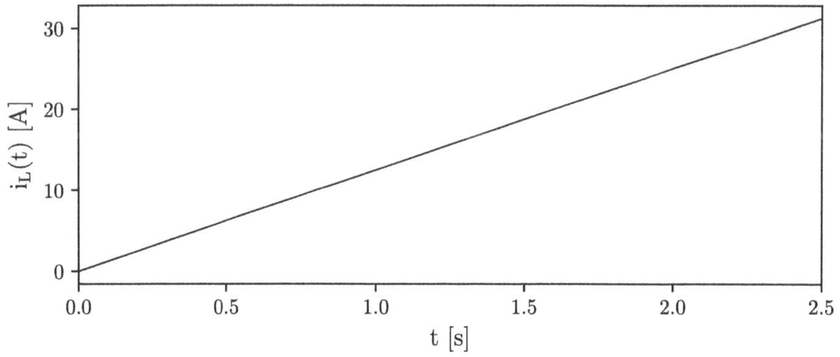

Fig. 1.152 Time evolution of $i_L(t)$

1.7.30 Obtaining the Differential Equation

In the circuit represented in Fig. 1.153, obtain the differential equation of the voltage $u(t)$ as a function of $i_g(t)$, R_1, R_2 and L.

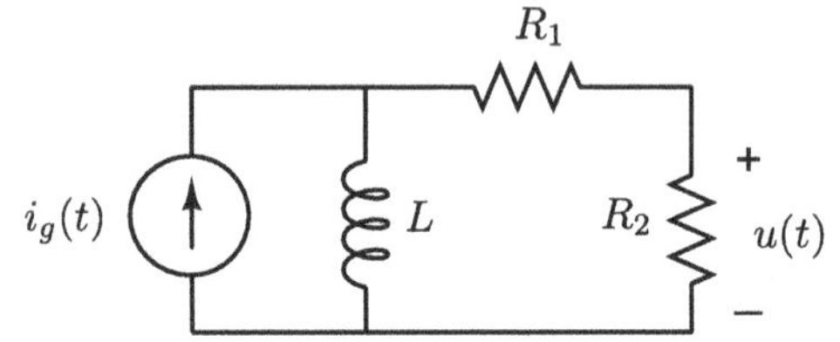

Fig. 1.153 Circuit considered in the problem under study

Solution. To simplify the notation, additional variables have been included, as shown in Fig. 1.154.

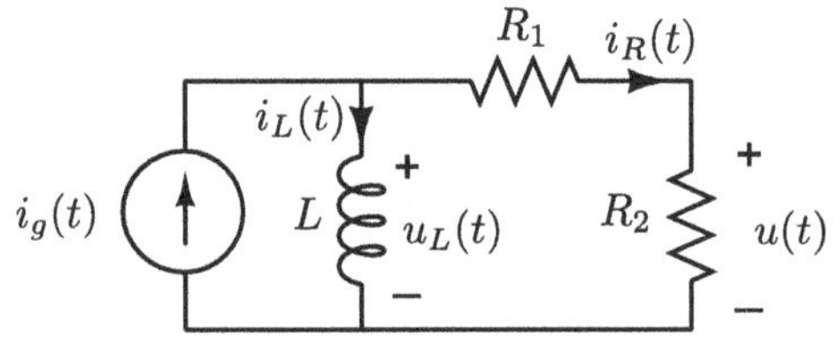

Fig. 1.154 Simplified circuit

Using Kirchhoff's current law:

$$i_g(t) = i_L(t) + i_R(t) \tag{1.51}$$

Given that

$$u(t) = R_2 \cdot i_R(t)$$

then (1.51) yields

$$i_g(t) = i_L(t) + \frac{u(t)}{R_2} \tag{1.52}$$

The differentiation of Eq. (1.52) results in:

$$\frac{di_g(t)}{dt} = \frac{di_L(t)}{dt} + \frac{1}{R_2} \cdot \frac{du(t)}{dt}$$

and considering that

$$u_L = L \cdot \frac{di_L(t)}{dt}$$

then

$$\frac{di_g(t)}{dt} = \frac{u_L(t)}{L} + \frac{1}{R_2} \cdot \frac{du(t)}{dt}$$

The relation between $u_L(t)$ and $u(t)$ is obtained by applying the concept of voltage divider:

$$u(t) = u_L(t) \cdot \frac{R_2}{R_1 + R_2} \quad \Rightarrow \quad u_L(t) = u(t) \cdot \frac{R_1 + R_2}{R_2}$$

Substituting and rearranging terms result in:

$$\frac{du(t)}{dt} + \frac{R_1 + R_2}{L} \cdot u(t) = R_2 \cdot \frac{di_g(t)}{dt}$$

1.7.31 Obtaining the Differential Equation

In the circuit represented in Fig. 1.155, deduce the differential equation of the current $i(t)$ as a function of $i_g(t)$, R_1, R_2 and L.

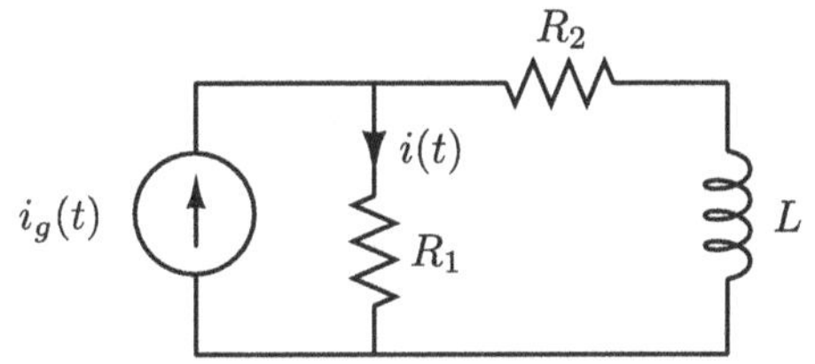

Fig. 1.155 Circuit considered in the problem under study

Solution. To simplify the notation, additional variables have been included, as shown in Fig. 1.156.

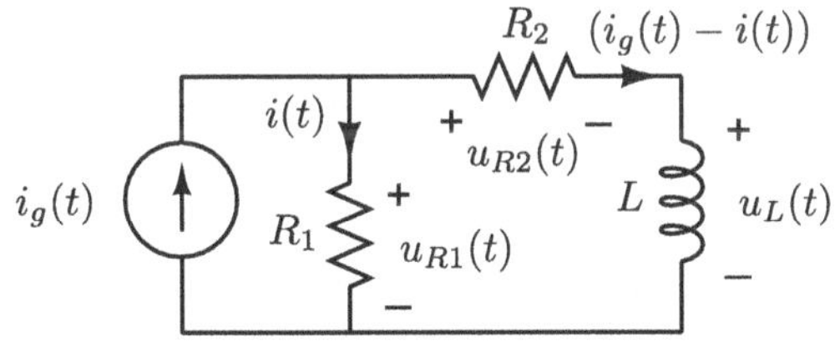

Fig. 1.156 Circuit with additional variables

Using Kirchhoff's voltage law in the loop formed by R_1, R_2 and L results in:

$$u_{R1}(t) = u_{R2}(t) + u_L(t) \tag{1.53}$$

Given that

$$u_L(t) = L \cdot \frac{d(i_g(t) - i(t))}{dt} \ ; \ u_{R1}(t) = R_1 \cdot i(t) \ ; \ u_{R2} = R_2 \cdot (i_g(t) - i(t))$$

then Eq. (1.53) yields

$$R_1 \cdot i(t) = R_2 \cdot (i_g(t) - i(t)) + L \cdot \frac{d(i_g(t) - i(t))}{dt}$$

Rearranging terms:

$$\frac{di(t)}{dt} + \frac{R_1 + R_2}{L} \cdot i(t) = \frac{di_g(t)}{dt} + \frac{R_2}{L} \cdot i_g(t)$$

Chapter 2
Second Order Transients

Abstract This chapter covers the second order circuits, beginning with a theoretical introduction of the concepts required to correctly address each of the subsequent problems. A total of 16 fully solved problems with explanatory comments are included.

2.1 Second Order Circuits

In second order circuits, any voltage or current can be obtained through a second order differential equation. Some examples of these circuits are:

- Circuits including two different types of energy storage elements, an inductor and a capacitors.

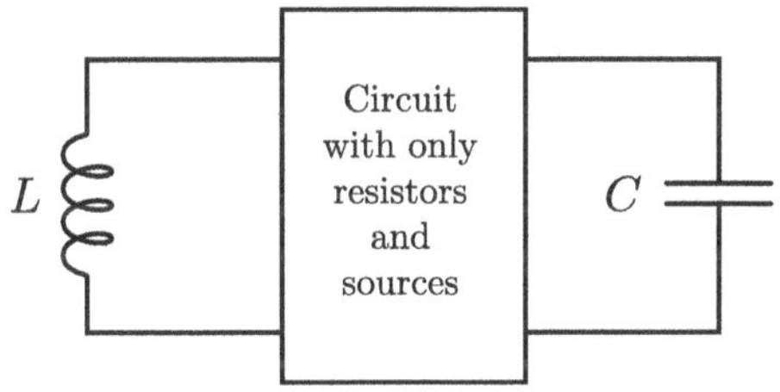

Fig. 2.1 Circuit with two different types of energy storage elements

- Circuits where there are two energy storage elements of the same type (inductor or capacitor) which cannot be reduced to a single equivalent.

Very different topologies can be found for second order circuits. This chapter focuses on the study of only two particular cases series RLC circuits and parallel RLC circuits, where it is relatively easy to determine the coefficients of the associated differential equation and, therefore, its resolution. The difficulty found in other second-order circuits to obtain the corresponding differential equation makes

A. Bachiller Soler et al., *Solved Problems for Transient Electrical Circuits*,
Lecture Notes in Electrical Engineering 809,
https://doi.org/10.1007/978-3-030-88144-3_2

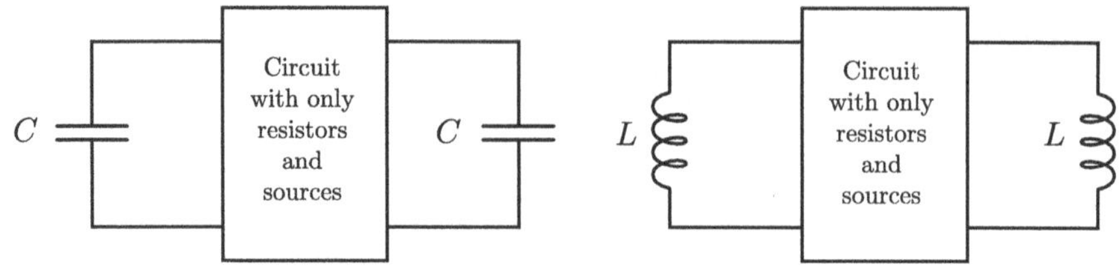

Fig. 2.2 Circuits with two energy storage elements of the same type

it advisable to use other resolution methods, such as state variable equations or Laplace transform.

2.1.1 Series RLC Circuit

First, the series RLC circuit under voltage supply will be discussed, as shown in Fig. 2.3.

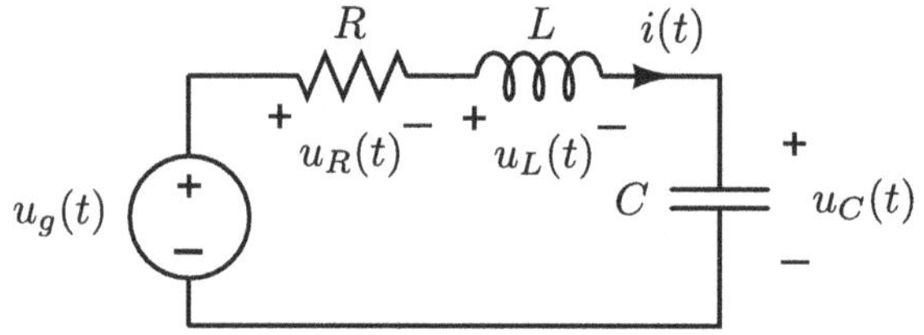

Fig. 2.3 Series RLC circuit

It can be noticed that many circuits have a topology similar to that represented in Fig. 2.3, including several inductors and capacitors which can be replaced by an equivalent inductor in series with an equivalent capacitor, the resulting structure is equivalent to that of a series RLC circuit. As an illustrative example, Fig. 2.4 shows the reduction of a circuit to its series RLC equivalent, where the static circuit has been replaced by its Thévenin equivalent and the energy storage elements by their corresponding equivalent elements.

The differential equation defining the behavior of the variables involved in the series RLC circuit represented in the Fig. 2.3 will be obtained in the sequel.

Differential equation of circuit current
Applying Kirchhoff's voltage law results in:

$$u_R(t) + u_L(t) + u_C(t) = u_g(t) \tag{2.1}$$

Considering Ohm's law in the resistor, jointly with the inductor defining equation:

$$u_R(t) = Ri(t) \quad ; \quad u_L(t) = L\frac{di(t)}{dt} \tag{2.2}$$

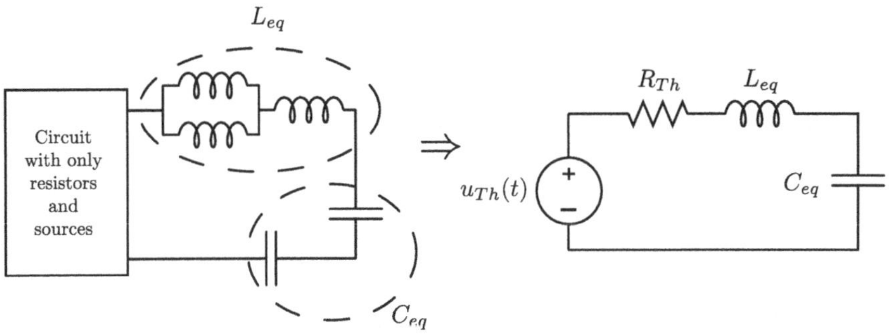

Fig. 2.4 Reduction to the equivalent series RLC circuit

so:

$$Ri(t) + L\frac{di(t)}{dt} + u_C(t) = u_g(t) \tag{2.3}$$

Differentiating the previous equation results in:

$$R\frac{di(t)}{dt} + L\frac{d^2i(t)}{dt^2} + \frac{du_C(t)}{dt} = \frac{du_g(t)}{dt} \tag{2.4}$$

and considering the capacitor defining equation

$$i(t) = C\frac{du_C(t)}{dt} \tag{2.5}$$

the following result is obtained:

$$R\frac{di(t)}{dt} + L\frac{d^2i(t)}{dt^2} + \frac{1}{C}i(t) = \frac{du_g(t)}{dt} \tag{2.6}$$

Finally, rearranging terms:

$$\frac{d^2i(t)}{dt^2} + \frac{R}{L}\frac{di(t)}{dt} + \frac{1}{LC}i(t) = \frac{1}{L}\frac{du_g(t)}{dt} \tag{2.7}$$

Differential equation of the capacitor voltage

In Eq. (2.3), the current is substituted by applying the capacitor defining equation (2.5), yielding:

$$RC\frac{du_C(t)}{dt} + LC\frac{d^2u_C(t)}{dt^2} + u_C(t) = u_g(t) \tag{2.8}$$

rearranging terms:

$$\frac{d^2u_C(t)}{dt^2} + \frac{R}{L}\frac{du_C(t)}{dt} + \frac{1}{LC}u_C(t) = \frac{1}{LC}u_g(t) \tag{2.9}$$

Differential equation of the resistor voltage
In Eq. (2.7), the current $i(t)$ can be replaced, according to Ohm's law, by $i(t){=}u_R(t)/R$, obtaining, after ordering terms:

$$\frac{d^2u_R(t)}{dt^2} + \frac{R}{L}\frac{du_R(t)}{dt} + \frac{1}{LC}u_R(t) = \frac{R}{L}\frac{du_g(t)}{dt} \tag{2.10}$$

Differential equation of the inductor voltage
Differentiating equation (2.1), the following result is obtained:

$$\frac{du_R(t)}{dt} + \frac{du_L(t)}{dt} + \frac{du_C(t)}{dt} = \frac{du_g(t)}{dt} \tag{2.11}$$

Considering Ohm's law in the resistor, jointly with the defining equation of the capacitor

$$u_R(t) = Ri(t) \quad ; \quad i(t) = C\frac{du_C(t)}{dt} \tag{2.12}$$

it results in:

$$R\frac{di(t)}{dt} + \frac{du_L(t)}{dt} + \frac{i(t)}{C} = \frac{du_g(t)}{dt} \tag{2.13}$$

Differentiating this equation again:

$$R\frac{d^2i(t)}{dt^2} + \frac{d^2u_L(t)}{dt^2} + \frac{1}{C}\frac{di(t)}{dt} = \frac{d^2u_g(t)}{dt^2} \tag{2.14}$$

Applying the inductor defining equation:

$$\frac{di(t)}{dt} = \frac{u_L(t)}{L} \tag{2.15}$$

then

$$\frac{R}{L}\frac{du_L(t)}{dt} + \frac{d^2u_L(t)}{dt^2} + \frac{1}{LC}u_L(t) = \frac{d^2u_g(t)}{dt^2} \tag{2.16}$$

Finally, rearranging terms:

$$\frac{d^2u_L(t)}{dt^2} + \frac{R}{L}\frac{du_L(t)}{dt} + \frac{1}{LC}u_L(t) = \frac{d^2u_g(t)}{dt^2} \tag{2.17}$$

In the differential equations obtained for the variables (2.7), (2.9), (2.10) and (2.17), it can be seen that the coefficients are equal in all of them, being the only difference in the independent term.

2.1.2 Parallel RLC Circuit

The parallel RLC circuit under current supply will be analyzed in this section, as shown in Fig. 2.5.

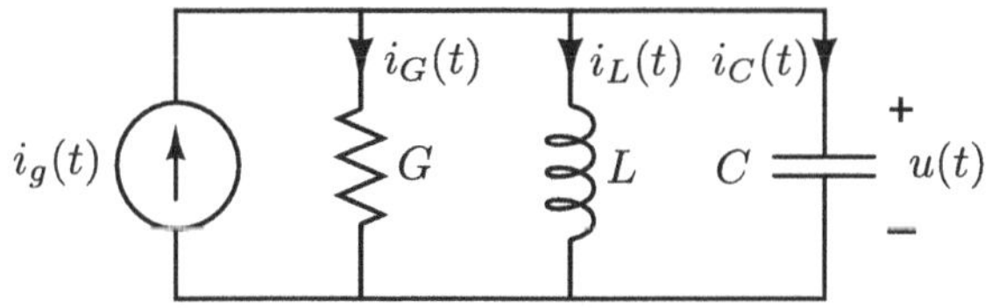

Fig. 2.5 Parallel RLC circuit

As mentioned for the series RLC circuit, it must be considered that many other circuits have the same topology, which can be reduced to that shown in Fig. 2.5. Thus, if in the circuit under study all the capacitors and inductors can be replaced by an equivalent capacitor in parallel with an equivalent inductor, being the rest of the circuit substituted by its Norton equivalent in order to have the parallel RLC shown in Fig. 2.6.

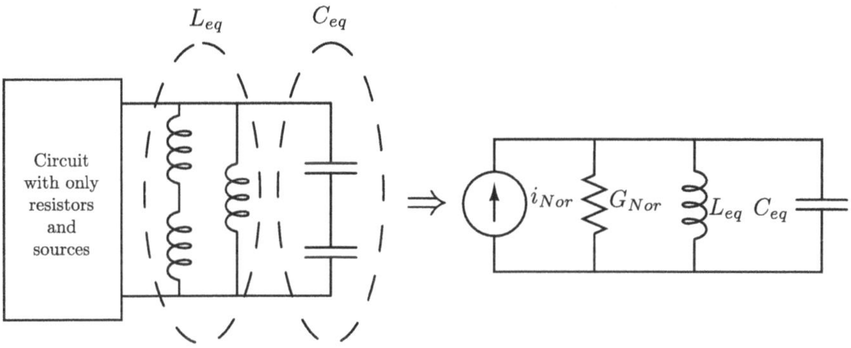

Fig. 2.6 Reduction to the equivalent parallel RLC circuit

The differential equation defining the behavior of the variables involved in the circuit represented in Fig. 2.5 will be obtained in the sequel.

Differential equation of circuit voltage
Applying Kirchhoff's current law results in:

$$i_G(t) + i_L(t) + i_C(t) = i_g(t) \tag{2.18}$$

Considering Ohm's Law in the resistor, jointly with the capacitor defining equation:

$$i_G(t) = Gu(t) \quad ; \quad i_C(t) = C\frac{du(t)}{dt} \tag{2.19}$$

then

$$Gu(t) + i_L(t) + C\frac{du(t)}{dt} = i_g(t) \tag{2.20}$$

Differentiating the previous equation results in:

$$G\frac{du(t)}{dt} + \frac{di_L(t)}{dt} + C\frac{d^2u(t)}{dt^2} = \frac{di_g(t)}{dt} \tag{2.21}$$

Using the defining equation of the inductor

$$u(t) = L\frac{di_L(t)}{dt} \tag{2.22}$$

the following expression is obtained:

$$G\frac{du(t)}{dt} + \frac{1}{L}u(t) + C\frac{d^2u(t)}{dt^2} = \frac{di_g(t)}{dt} \tag{2.23}$$

Finally, rearranging terms:

$$\frac{d^2u(t)}{dt^2} + \frac{G}{C}\frac{du(t)}{dt} + \frac{1}{LC}u(t) = \frac{1}{C}\frac{di_g(t)}{dt} \tag{2.24}$$

Differential equation of inductor current
In Eq. (2.20), the voltage is substituted by applying the inductor defining equation (2.22), yielding:

$$GL\frac{di_L(t)}{dt} + i_L(t) + LC\frac{d^2i_L(t)}{dt^2} = i_g(t) \tag{2.25}$$

and rearranging terms:

$$\frac{d^2i_L(t)}{dt^2} + \frac{G}{C}\frac{di_L(t)}{dt} + \frac{1}{LC}i_L(t) = \frac{1}{LC}i_g(t) \tag{2.26}$$

Differential equation of resistor current
In Eq. (2.24), the voltage $u(t)$ can be replaced, according to Ohm's law, by $u(t)=i_G(t)/G$, obtaining, after ordering terms:

$$\frac{d^2i_G(t)}{dt^2} + \frac{G}{C}\frac{di_G(t)}{dt} + \frac{1}{LC}i_G(t) = \frac{G}{C}\frac{di_g(t)}{dt} \tag{2.27}$$

Differential equation of the inductor voltage
Differentiating equation (2.18), the following expression is obtained:

$$\frac{di_G(t)}{dt} + \frac{di_L(t)}{dt} + \frac{di_C(t)}{dt} = \frac{di_g(t)}{dt} \tag{2.28}$$

Considering Ohm's law in the resistor, jointly with the defining equation of the inductor

$$i_G(t) = Gu(t) \quad ; \quad u(t) = L\frac{di_L(t)}{dt} \tag{2.29}$$

results in:

$$G\frac{du(t)}{dt} + \frac{u(t)}{L} + \frac{di_C(t)}{dt} = \frac{di_g(t)}{dt} \tag{2.30}$$

Differentiating the previous equation again:

$$G\frac{d^2u(t)}{dt^2} + \frac{1}{L}\frac{du(t)}{dt} + \frac{d^2i_C(t)}{dt^2} = \frac{d^2i_g(t)}{dt^2} \tag{2.31}$$

Using the capacitor defining equation

$$\frac{du(t)}{dt} = \frac{i_C(t)}{C} \tag{2.32}$$

then:

$$\frac{G}{C}\frac{di_C(t)}{dt} + \frac{1}{LC}i_C(t) + \frac{d^2i_C(t)}{dt^2} = \frac{d^2i_g(t)}{dt^2} \tag{2.33}$$

Finally, rearranging terms:

$$\frac{d^2i_C(t)}{dt^2} + \frac{G}{C}\frac{di_C(t)}{dt} + \frac{1}{LC}i_C(t) = \frac{d^2i_g(t)}{dt^2} \tag{2.34}$$

In the differential equations obtained for the variables (2.24), (2.26), (2.27) and (2.34), it can be noticed that the coefficients are equal in all of them, being the only difference in the independent term. This fact will be used to write a generic equation for all the variables.

2.1.3 *Generic Differential Equation of a Second Order Circuit*

In Fig. 2.7, the equations of all the variables of the series and parallel RLC circuits are shown as they were obtained in the previous sections.

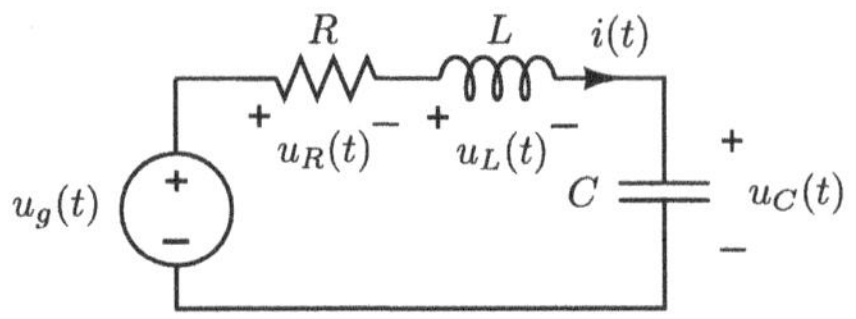

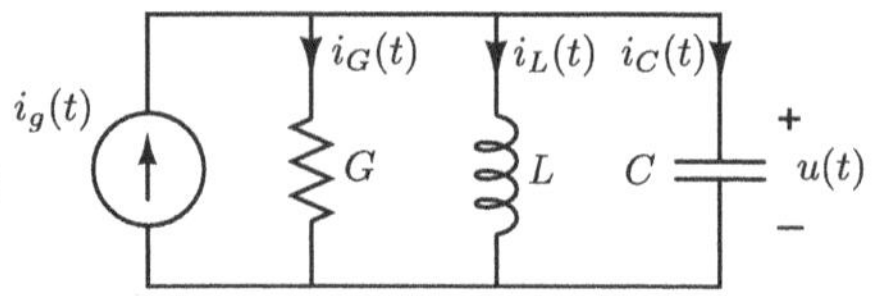

$$\frac{d^2i(t)}{dt^2}+\frac{R}{L}\frac{di(t)}{dt}+\frac{1}{LC}i(t)=\frac{1}{L}\frac{du_g(t)}{dt}$$

$$\frac{d^2u_C(t)}{dt^2}+\frac{R}{L}\frac{du_C(t)}{dt}+\frac{1}{LC}u_C(t)=\frac{u_g(t)}{LC}$$

$$\frac{d^2u_R(t)}{dt^2}+\frac{R}{L}\frac{du_R(t)}{dt}+\frac{1}{LC}u_R(t)=\frac{R}{L}\frac{du_g(t)}{dt}$$

$$\frac{d^2u_L(t)}{dt^2}+\frac{R}{L}\frac{du_L(t)}{dt}+\frac{1}{LC}u_L(t)=\frac{d^2u_g(t)}{dt^2}$$

$$\frac{d^2u(t)}{dt^2}+\frac{G}{C}\frac{du(t)}{dt}+\frac{1}{LC}u(t)=\frac{1}{C}\frac{di_g(t)}{dt}$$

$$\frac{d^2i_L(t)}{dt^2}+\frac{G}{C}\frac{di_L(t)}{dt}+\frac{1}{LC}i_L(t)=\frac{i_g(t)}{LC}$$

$$\frac{d^2i_G(t)}{dt^2}+\frac{G}{C}\frac{di_G(t)}{dt}+\frac{1}{LC}i_G(t)=\frac{G}{C}\frac{di_g(t)}{dt}$$

$$\frac{d^2i_C(t)}{dt^2}+\frac{G}{C}\frac{di_C(t)}{dt}+\frac{1}{LC}i_C(t)=\frac{d^2i_g(t)}{dt^2}$$

Fig. 2.7 Differential equations of series and parallel RLC circuits

According to Fig. 2.7, it can be noticed that all equations can be expressed in the following generic form

$$\frac{d^2 f(t)}{dt^2}+2\alpha\frac{df(t)}{dt}+\omega_0^2 f(t)=g(t) \tag{2.35}$$

where $f(t)$ denotes the voltage or current to be determined, $g(t)$ is a function related to the excitation source of the circuit, and α and ω_0 are constants that depend on the parameters of the passive elements.

The constant α is called **damping coefficient**, being its SI unit s^{-1}. This constant is characteristic of each circuit and its value is given by:

- Series RLC circuit:

$$\alpha=\frac{R}{2L} \tag{2.36}$$

- Parallel RLC circuit:

$$\alpha=\frac{G}{2C} \tag{2.37}$$

On the other hand, the constant ω_0 is called the **resonant frequency**, being its SI unit rad s^{-1}. This parameter is also characteristic of each circuit, taking the same value for the series and parallel RLC circuits:

$$\omega_0=\frac{1}{\sqrt{LC}} \tag{2.38}$$

It is easy to calculate the values of these constants for circuits which can be transformed into a series or parallel RLC circuit, where, in order to obtain α and ω_0, the values of R_{eq}, G_{eq}, L_{eq} and C_{eq} will be considered. In any other type of second-order circuit, the determination of the constants α and ω_0 requires setting up the differential equation, which in some cases can be a laborious process. In those cases it is preferred to use other resolution methods, such as the Laplace transform.

2.2 Transient Response of Second Order Circuits

As explained before, all the variables of a second order circuit are given by the following constant-coefficient linear differential equation:

$$\frac{d^2 f(t)}{dt^2} + 2\alpha \frac{df(t)}{dt} + \omega_0^2 f(t) = g(t) \tag{2.39}$$

The transient response of the variable considered will be obtained from the solution of this equation. The transient will be assumed to start at t=0, but it can easily be generalized to any other time t=t_0. As in first-order circuits, the complete response is made up of two terms: the natural response, $f_n(t)$, solution of the homogeneous equation, and the forced or steady-state response, $f_p(t)$, corresponding to a particular solution:

$$f(t) = f_n(t) + f_p(t) \tag{2.40}$$

2.2.1 *Natural Response*

The natural response is the solution of the homogeneous second-order differential equation:

$$\frac{d^2 f(t)}{dt^2} + 2\alpha \frac{df(t)}{dt} + \omega_0^2 f(t) = 0 \tag{2.41}$$

For this resolution, the associated characteristic equation will be obtained, using the Laplace transform with zero initial values (see chapter 3):

$$s^2 + 2\alpha \cdot s + \omega_o^2 = 0 \tag{2.42}$$

being the solutions of this equation denoted as natural frequencies of the circuit:

$$s_1, s_2 = -\alpha \pm \sqrt{\alpha^2 - \omega_o^2} \tag{2.43}$$

Depending on the values of the constants α and ω_0, the natural frequencies can be either real or complex, meaning that the natural response has different time evolution. Three cases are described below:

- **Overdamped natural response**. This type of response appears when the damping coefficient is greater than the resonant frequency ($\alpha > \omega_0$), resulting in real and negative values of the natural frequencies, s_1 and s_2:

$$s_1,s_2 = -\alpha \pm \sqrt{\alpha^2 - \omega_o^2} \in \mathbb{R},\ s_1,s_2 < 0 \tag{2.44}$$

being the natural response the sum of two real exponential functions:

$$f_n(t) = K_1 \cdot e^{s_1 t} + K_2 \cdot e^{s_2 t} \tag{2.45}$$

- **Critically damped natural response**. If the damping coefficient and the resonant frequency have the same value ($\alpha = \omega_0$), an unique (real) value of the natural frequency is obtained:

$$s_1 = s_2 = -\alpha < 0 \tag{2.46}$$

The natural response in this case yields:

$$f_n(t) = (K_1 \cdot t + K_2) \cdot e^{-\alpha t} \tag{2.47}$$

- **Underdamped natural response** Finally, when $\alpha < \omega_0$, the natural frequencies of the circuit, s_1 and s_2, are complex conjugate:

$$s_1,s_2 = -\alpha \pm j\sqrt{\omega_o^2 - \alpha^2} = -\alpha \pm j\omega_d \in \mathbb{C} \tag{2.48}$$

The natural response is constituted by the sum of two conjugated complex exponential functions, although it is preferable to express it as a damped sinusoidal wave:

$$f_n(t) = K \cdot e^{-\alpha t} \cdot \sin(\omega_d t + \varphi) \tag{2.49}$$

where ω_d is called damped frequency.
A particular case of the underdamped response is the so-called **no damping** or no losses, which is produced when the damping coefficient is zero (α=0). The natural frequencies are in this case

$$s_1,s_2 = \pm j\omega_0 \in \mathbb{C} \tag{2.50}$$

and the natural response is a pure sinusoidal wave of frequency ω_0:

$$f_n(t) = K \cdot \sin(\omega_0 t + \varphi) \tag{2.51}$$

The different forms that the time evolution of the natural response acquires according to the damping have been represented in Fig. 2.8. The underdamped response is characterized by its oscillations around the final value, being the amplitude of the sine wave modulated by a decreasing exponential. In contrast, the overdamped response tends asymptotically to the final value. The critically damped case, the limit between

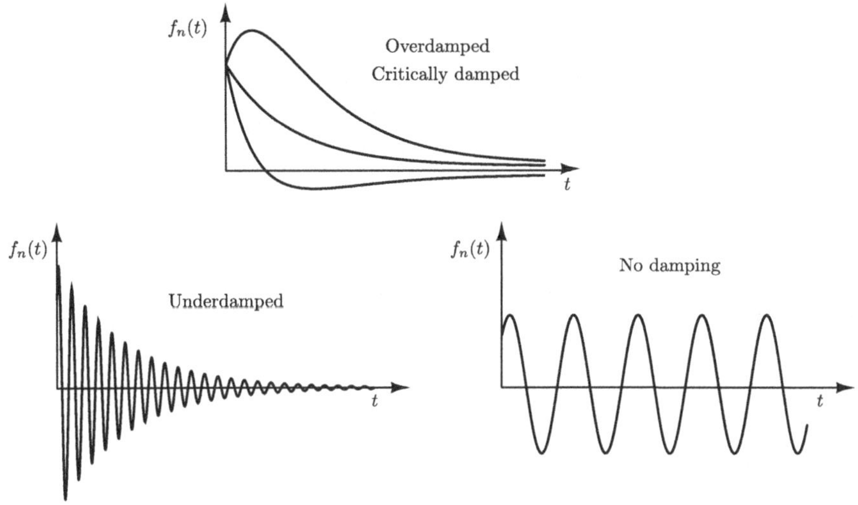

Fig. 2.8 Time evolution of the natural response according to damping

both types of responses, does not present oscillations and cannot be graphically distinguished from the overdamped case. Finally, for a circuit with no damping, the natural response is a sinusoidal wave whose amplitude remains constant over time.

2.2.2 Forced or Steady State Response

The forced response corresponds mathematically to a particular solution of the complete differential equation. Although it can be obtained by mathematical methods without taking into account the electrical nature of the equation, it is easier to determine it with specific techniques used in steady-state analysis of electrical circuits with dc or ac excitation. So, for dc-type sources, the steady-state dc circuit will be solved considering the behavior of the capacitor (open circuit) and the inductor (short circuit) in this regime. On the other hand, for ac-type sources, the steady-state ac circuit will be obtained in the phasor domain and consequently solved.

2.2.3 Complete Response

Once the natural and the steady-state responses have been determined, the complete response can be obtained as the sum of both terms:

$$f(t) = f_n(t) + f_p(t) \tag{2.52}$$

Depending on the type of natural response, the complete response will be:

$$f(t) = K_1 \cdot e^{s_1 t} + K_2 \cdot e^{s_2 t} + f_p(t) \tag{2.53}$$

$$f(t) = (K_1 \cdot t + K_2) \cdot e^{-\alpha t} + f_p(t) \tag{2.54}$$

$$f(t) = K \cdot e^{-\alpha t} \cdot \sin(\omega_d t + \varphi) + f_p(t) \tag{2.55}$$

Once the response in steady state, $f_p(t)$, has been obtained, jointly with the type of natural response, the values of the different constants (K_1, K_2) or (K, φ) can be calculated in order to fully define the complete response of the variable under study. These constants are obtained from the value at instant t=0^+ of the corresponding variable, $f(0^+)$, and its first derivative, $f'(0^+)$.

2.2.4 *Initial Conditions*

As indicated before, during the transient period there is a redistribution of energy, which cannot take place instantaneously, meaning that, in absence of impulse-type responses, it can be verified that:

1. The capacitor voltage cannot suffer discontinuities:

$$u_C(0^+) = u_C(0^-) \tag{2.56}$$

2. The inductor current cannot suffer discontinuities:

$$i_L(0^+) = i_L(0^-) \tag{2.57}$$

With these premises, the initial value of any voltage or current, $f(0^+)$, and its derivative, $f'(0^+)$ can be calculated. For this purpose, two different circuits are considered: a first one where the initial value of the variable is determined and a second circuit for the initial value of the derivative.

Obtaining the initial condition of the variable: $f(0^+)$
The initial value of the variable considered is obtained by solving a circuit where:

1. The excitation sources, $e_g(t)$ and $i_g(t)$, are replaced by two sources with constant values equal to:

$$E_g = e_g(0^+) \ ; \ \ I_g = i_g(0^+) \tag{2.58}$$

2. The capacitor is replaced by a voltage source of value:

$$u_C(0^+) = u_C(0^-) = U_0 \tag{2.59}$$

3. The inductor is replaced by a current source of value:

$$i_L(0^+) = i_L(0^-) = I_0 \tag{2.60}$$

The process to obtain the circuit has been synthesized in Fig. 2.9 at t=0^+, this scheme allows obtaining the initial value of any voltage or current, $f(0^+)$.

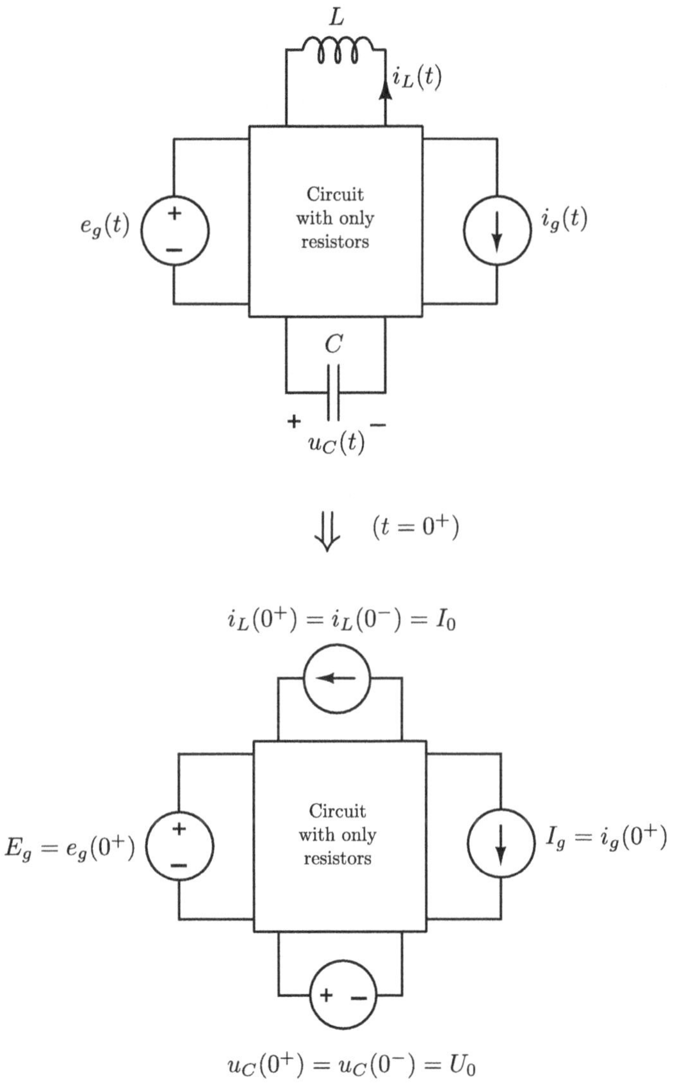

Fig. 2.9 Obtaining the circuit at t=0^+

Obtaining the initial condition of the derivative: $f'(0^+)$
To simplify the notation, the following expressions will be equivalent:

$$f'(0^+) \triangleq \left.\frac{df(t)}{dt}\right|_{t=0^+} \triangleq \frac{df(0^+)}{dt} \tag{2.61}$$

For any linear circuit, if a set of voltages verify Kirchhoff's voltage law, the derivative of these voltages also fulfill this equality. The same is verified for Kirchhoff's current law, with Ohm's law and with the defining equation of the inductor and the capacitor. In summary, the same voltage and current equations in a circuit are also fulfilled by the corresponding derivatives, which allows obtaining the derivative of any voltage or current at instant t=0^+ solving a circuit where:

1. The excitation sources, $e_g(t)$ and $i_g(t)$, are replaced by two sources with constant values equal to:
$$\frac{de_g(0^+)}{dt} \;;\; \frac{di_g(0^+)}{dt} \tag{2.62}$$
2. The capacitor is replaced by a voltage source of value:
$$\frac{du_C(0^+)}{dt} = \frac{i_C(0^+)}{C} \tag{2.63}$$
3. The inductor is replaced by a current source of value:
$$\frac{di_L(0^+)}{dt} = \frac{u_L(0^+)}{L} \tag{2.64}$$

Figure 2.10 represents the procedure used to obtain the "derivative" circuit at t=0^+ that allows calculating the initial value of the derivative of any voltage or current, $f'0^+)$. It can be seen that this circuit has the same topology as that in Fig. 2.9, with the derivatives instead of the original variables.

2.3 Procedure to Obtain the Response of a Second Order Circuit

As a summary, the following steps are used to obtain the voltage or current of any element in a second order circuit during a transient starting at t=0^+. If the variable to be calculated is denoted by $f(t)$, the complete response is given by:

$$f(t) = f_n(t) + f_p(t) \tag{2.65}$$

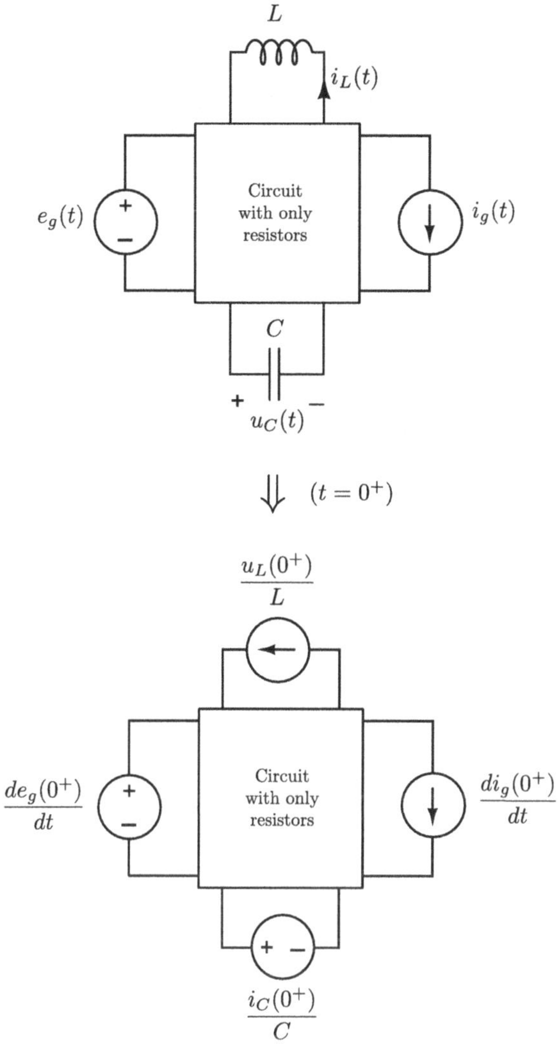

Fig. 2.10 Obtaining the "derivative" circuit at t=0^+

The process to obtain $f(t)$ is:

1. Determine the steady-state response, $f_p(t)$ for $t \geq 0$. For this purpose, any of the known techniques for steady-state circuit analysis in dc or ac can be used, depending on the sources involved. If the circuit does not have independent excitation sources for $t \geq 0$, $f_p(t)$ will be zero.
2. Determine the natural response of the circuit. To do this, the damping coefficient and the resonant frequency will be calculated depending on the type of circuit:

   - Series RLC circuit or equivalent. If the independent sources are canceled and the resultant passive circuit is a series RLC, then:

$$\alpha = \frac{R}{2L} \quad ; \quad \omega_0 = \frac{1}{\sqrt{LC}}$$

- Parallel RLC circuit or equivalent. If the independent sources are canceled and the resultant passive circuit is a parallel RLC, then:

$$\alpha = \frac{G}{2C} \quad ; \quad \omega_0 = \frac{1}{\sqrt{LC}}$$

- Another type of second order circuit. In this case, it is not possible to know a priori the values of α and ω_0, so it is necessary to determine the differential equation of some variable in the circuit to calculate them. Alternatively, other methods can be used to solve circuits in transient regime.

With the values of α and ω_0 the natural frequencies of the circuit, s_1 and s_2, are obtained:

$$s_1, s_2 = -\alpha \pm \sqrt{\alpha^2 - \omega_o^2}$$

and, according to its values, the type of natural response is obtained:

- **Overdamped** ($\alpha > \omega_0$): s_1,s_2=$-\alpha \pm \sqrt{\alpha^2 - \omega_o^2} \in \mathbb{R}$, s_1,s_2<0

$$f_n(t) = K_1 \cdot \mathrm{e}^{s_1 t} + K_2 \cdot \mathrm{e}^{s_2 t}$$

- **Critically damped** ($\alpha = \omega_0$): s_1=s_2=$-\alpha$<0

$$f_n(t) = (K_1 \cdot t + K_2) \cdot \mathrm{e}^{-\alpha t}$$

- **Underdamped** ($\alpha < \omega_0$): s_1,s_2=$-\alpha \pm j\sqrt{\omega_o^2 - \alpha^2}$=$-\alpha \pm j\omega_d \in \mathbb{C}$

$$f_n(t) = K \cdot \mathrm{e}^{-\alpha t} \cdot \sin(\omega_d t + \varphi)$$

3. Determine the initial value of the variable, $f(0^+)$, and that of its derivative $f'(0^+)$. To do this, the circuits in Figs. 2.11 and 2.12 must be solved.
4. Determination of the constants in the complete response. Depending on the natural response type, $f(t)$ will contain a different pair of constants (K_1, K_2) or (K, φ). To calculate these parameters, it will be enough to substitute t=0 in the complete response and its derivative, and make them equal to the values obtained in the previous step, $f(0^+)$ and $f'(0^+)$, that is:

$$\left.\begin{aligned} f(0^+) &= f_n(0^+) + f_p(0^+) \\ f'(0^+) &= f_n'(0^+) + f_p'(0^+) \end{aligned}\right\} \Rightarrow \begin{matrix} (K_1, K_2) \\ \text{or} \\ (K, \varphi) \end{matrix}$$

It should be remarked that the first three steps are independent, so that the order can be freely changed. However, step number 4 should always be the last one, as it requires the previous results.

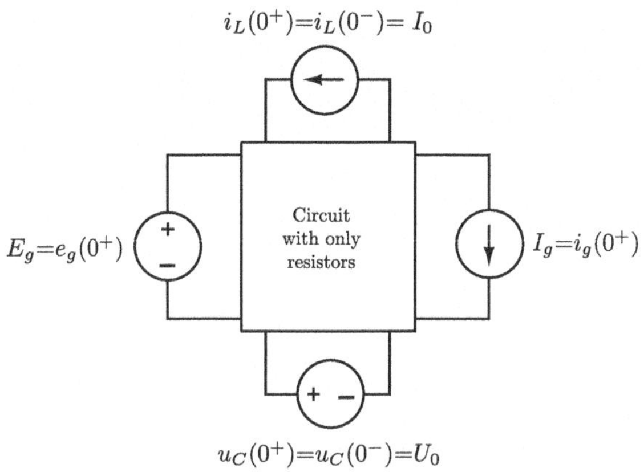

Fig. 2.11 Circuit to obtain $f(0^+)$

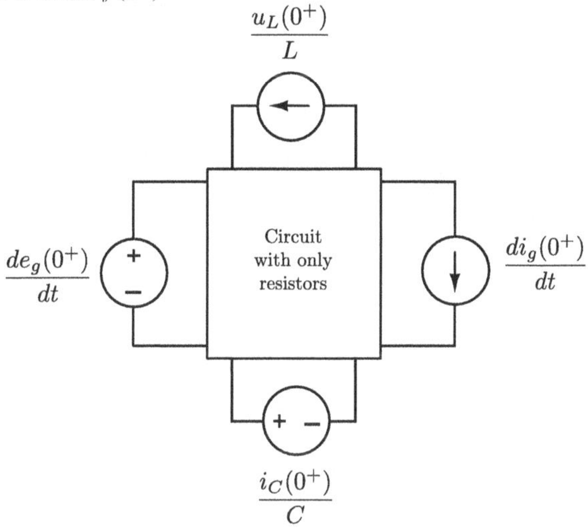

Fig. 2.12 Circuit to obtain $f'(0^+)$

2.4 Solved Problems

2.4.1 Series RLC. Calculation of Initial Conditions

In the circuit represented in Fig. 2.13, given that $i_L(0^-)=3\,\text{A}$ and $u_C(0^-)=2\,\text{V}$, calculate $i_L(0^+)$, $u_C(0^+)$, $u_L(0^+)$ and $u_R(0^+)$ and their derivatives $i'_L(0^+)$, $u'_C(0^+)$, $u'_L(0^+)$ and $u'_R(0^+)$.

Solution. To calculate the values of the magnitudes under study at $t=0^+$, the circuit will be analyzed at that instant, substituting the capacitor for a voltage source of

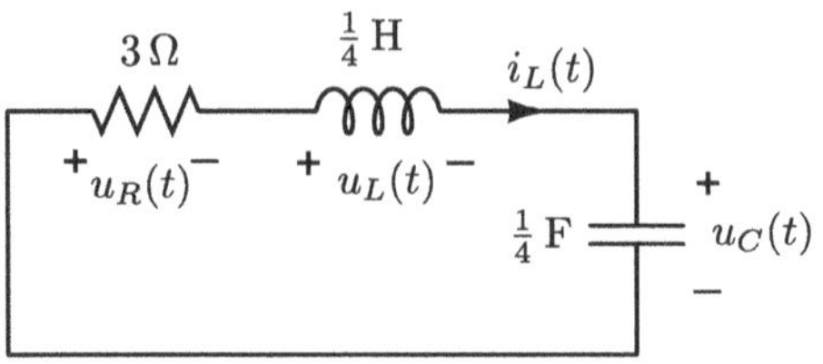

Fig. 2.13 Circuit considered in the problem under study

value $u_C(0^+)$, and the inductor by a current source of value $i_L(0^+)$. Except for an impulse-type response, the capacitor voltage and the current through the inductor do not change from t=0^- to t=0^+. Therefore:

$$i_L(0^+) = i_L(0^-) = 3\,\text{A} \ ; \ \ u_C(0^+) = u_C(0^-) = 2\,\text{V}$$

This way, the circuit at t=0^+ is that shown in Fig. 2.14.

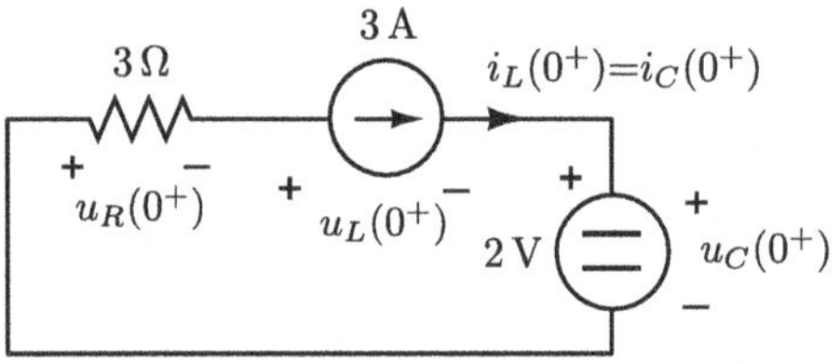

Fig. 2.14 Circuit at t=0^+

Analyzing the circuit depicted in Fig. 2.14, the rest of the variables are obtained at t=0^+ yield:

$$u_R(0^+) = 3 \cdot 3 = 9\,\text{V}$$
$$u_L(0^+) = -u_R(0^+) - u_C(0^+) = -9 - 2 = -11\,\text{V}$$

To calculate the derivative of these magnitudes at t=0^+, the derivative circuit will be analyzed, which is obtained by replacing the capacitor with a voltage source of value $u'_C(0^+)$ and the inductor with a current source of value $i'_L(0^+)$. These derivatives functions are obtained from the quantities calculated in the circuit at t=0^+ as follows:

$$u'_C(0^+) = \frac{i_C(0^+)}{C} = \frac{i_L(0^+)}{C} = \frac{3}{1/4} = 12\,\text{V/s}$$
$$i'_L(0^+) = \frac{u_L(0^+)}{L} = \frac{-11}{1/4} = -44\,\text{A/s}$$

Thus, the derivative circuit at t=0^+ is that shown in Fig. 2.15.

From the circuit in Fig. 2.15, the rest of the derivative variables are obtained at t=0$^+$:

$$u'_R(0^+) = 3 \cdot (-44) = -132\,\text{V/s}$$
$$u'_L(0^+) = -u'_R(0^+) - u'_C(0^+) = 132 - 12 = 120\,\text{V/s}$$

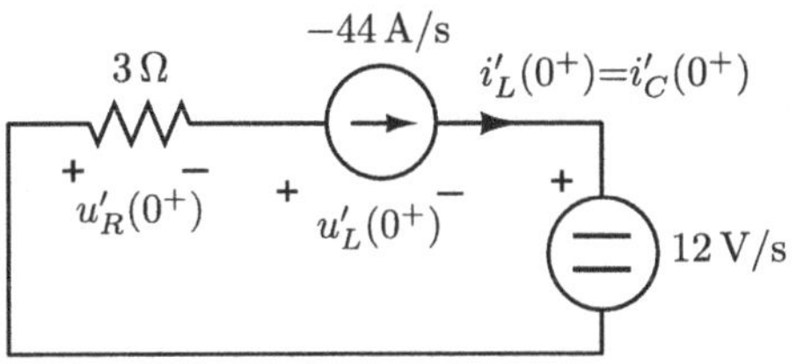

Fig. 2.15 Derivative circuit at t=0$^+$

2.4.2 Series RLC Without Excitation Sources

In the circuit represented in Fig. 2.16, given that $i_L(0^-)$=3 A and $u_C(0^-)$=2 V, calculate $u_C(t)$ for t>0 in the following cases:

(a) R=3 Ω
(b) R=2 Ω
(c) R=1 Ω
(d) R=0 Ω

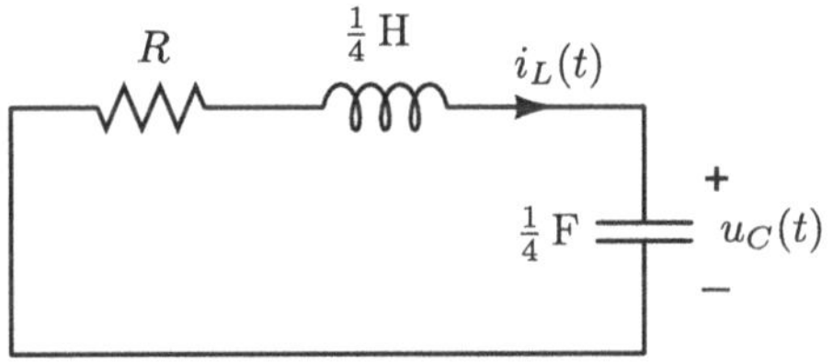

Fig. 2.16 Circuit considered in the problem under study

Solution. The response of a second-order circuit is obtained as the sum of the steady-state response and the natural response. Since the circuit has no excitation sources, the steady-state response is zero.

The natural response in a second order circuit is characterized by the value of the damping coefficient, α, and that of the resonant frequency, ω_0. Being a series RLC circuit, the values of these parameters are obtained based on R, L and C according to:

$$\alpha = \frac{R}{2L} \ ; \ \omega_0 = \frac{1}{\sqrt{LC}}$$

The initial conditions, $u_C(0^+)$ and $u'_C(0^+)$, are calculated by analyzing the circuit at t=0$^+$, which is obtained substituting the capacitor for a voltage source of value

$u_C(0^+)$ and the inductor for a current source of value $i_L(0^+)$. Except for an impulse-type response, the capacitor voltage and the inductor current do not change from $t{=}0^-$ to $t{=}0^+$. Therefore:

$$i_L(0^+) = i_L(0^-) = 3\,\text{A} \;\; ; \;\; u_C(0^+) = u_C(0^-) = 2\,\text{V}$$

This way, the circuit at $t{=}0^+$ is that shown in Fig. 2.17.

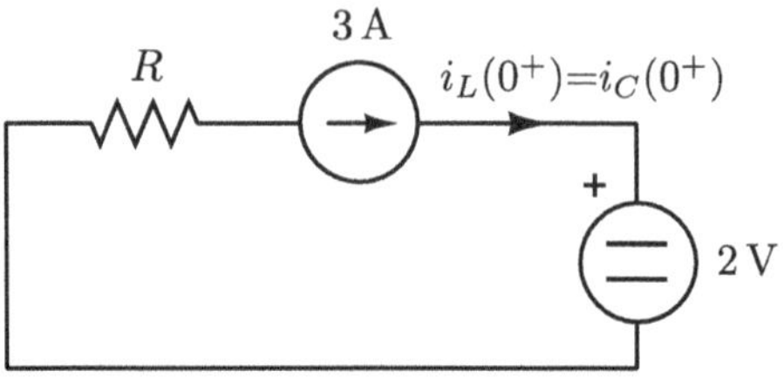

Fig. 2.17 Circuit at $t{=}0^+$

Analyzing the circuit in Fig. 2.17, $i_C(0^+)$ is obtained and, consequently, the value of $u'_C(0^+)$:

$$u'_C(0^+) = \frac{i_C(0^+)}{C} = \frac{3}{1/4} = 12\,\text{V/s}$$

Since both $u_C(0^+)$ and $u'_C(0^+)$ are independent of R, their values will be valid for all the case studies.

(a) For $R{=}3\,\Omega$, the values of α and ω_0 are calculated as:

$$\alpha = \frac{R}{2L} = \frac{3}{2\cdot(1/4)} = 6\,\text{s}^{-1} \;\; ; \;\; \omega_0 = \frac{1}{\sqrt{LC}} = \frac{1}{\sqrt{(1/4)\cdot(1/4)}} = 4\,\text{rad}\,\text{s}^{-1}$$

Since $\alpha{>}\omega_0$, the natural response is overdamped:

$$u_C^n(t) = K_1 \cdot \text{e}^{s_1 t} + K_2 \cdot \text{e}^{s_2 t}$$

where

$$s_1 = -\alpha + \sqrt{\alpha^2 - \omega_0^2} = -6 + \sqrt{6^2 - 4^2} \approx -1{,}53\,\text{s}^{-1}$$
$$s_2 = -\alpha - \sqrt{\alpha^2 - \omega_0^2} = -6 - \sqrt{6^2 - 4^2} \approx -10{,}47\,\text{s}^{-1}$$

Since there is no steady-state response:

$$u_C(t) = u_C^n(t) = K_1 \cdot \text{e}^{-1{,}53t} + K_2 \cdot \text{e}^{-10{,}47t}\,\text{V}$$

The values of the constants K_1 and K_2 are calculated from the initial conditions:

$$\left.\begin{aligned} u_C(0^+) &= 2 = K_1 + K_2 \\ u'_C(0^+) &= 12 = -1{,}53K_1 - 10{,}47K_2 \end{aligned}\right\} \Longrightarrow K_1 \approx 3{,}68\,\mathrm{V} \ ; \ K_2 \approx -1{,}68\,\mathrm{V}$$

Resulting in:

$$u_C(t) = 3{,}68 \cdot \mathrm{e}^{-1{,}53t} - 1{,}68 \cdot \mathrm{e}^{-10{,}47t}\ \mathrm{V}$$

The time evolution of $u_C(t)$ for R=3 Ω has been represented in Fig. 2.18.

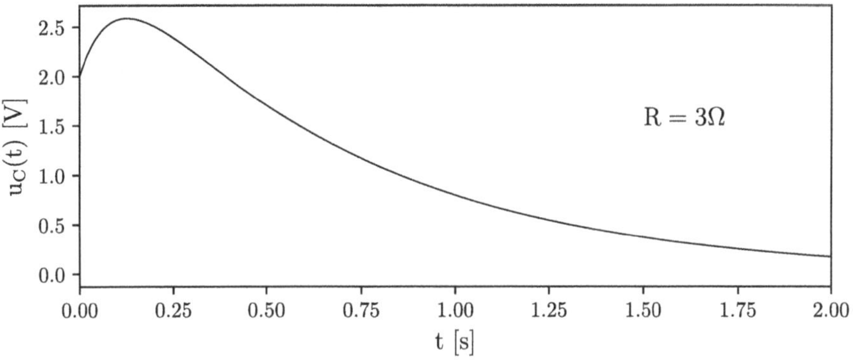

Fig. 2.18 Time evolution of $u_C(t)$ for R=3 Ω

(b) For R=2 Ω, the values of α and ω_0 are:

$$\alpha = \frac{R}{2L} = \frac{2}{2\cdot(1/4)} = 4\,\mathrm{s}^{-1} \ ; \ \omega_0 = \frac{1}{\sqrt{LC}} = \frac{1}{\sqrt{(1/4)\cdot(1/4)}} = 4\,\mathrm{rad\,s}^{-1}$$

Since α=ω_0, the natural response is critically damped:

$$u_C^n(t) = (K_1 + K_2 \cdot t)\cdot \mathrm{e}^{-\alpha t}$$

Since steady-state there is no response:

$$u_C(t) = u_C^n(t) = (K_1 + K_2 \cdot t)\cdot \mathrm{e}^{-4t}\ \mathrm{V}$$

The values of the constants K_1 and K_2 are calculated from the initial conditions:

$$\left.\begin{aligned} u_C(0^+) &= 2 = K_1 \\ u'_C(0^+) &= 12 = -4K_1 + K_2 \end{aligned}\right\} \Longrightarrow K_1 = 2\,\mathrm{V} \ ; \ K_2 = 20\,\mathrm{V/s}$$

Resulting in:

$$u_C(t) = (2 + 20\cdot t)\cdot \mathrm{e}^{-4t}\ \mathrm{V}$$

The time evolution of $u_C(t)$ for R=2 Ω has been represented in Fig. 2.19.

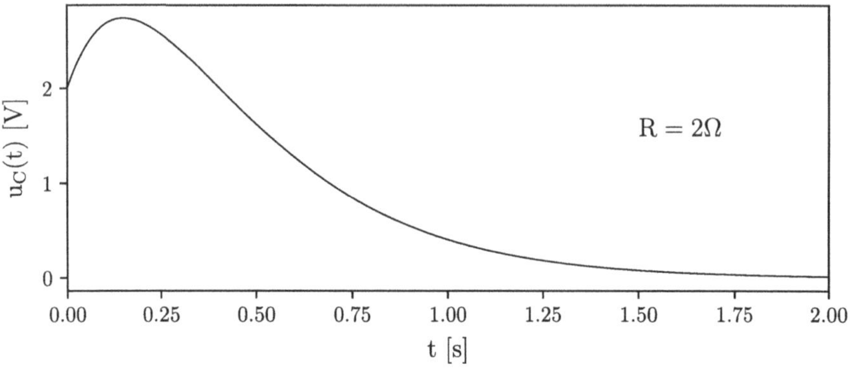

Fig. 2.19 Time evolution of $u_C(t)$ for R=2 Ω

(c) For R=1 Ω, the values of α and ω_0 are:

$$\alpha = \frac{R}{2L} = \frac{1}{2 \cdot (1/4)} = 2\,\mathrm{s}^{-1} \;\; ; \;\; \omega_0 = \frac{1}{\sqrt{LC}} = \frac{1}{\sqrt{(1/4) \cdot (1/4)}} = 4\,\mathrm{rad\,s}^{-1}$$

Since $\alpha < \omega_0$, the natural response is underdamped:

$$u_C^n(t) = K\mathrm{e}^{-\alpha t} \sin\left(\omega_d t + \varphi\right)$$

where:

$$\omega_d = \sqrt{\omega_0^2 - \alpha^2} = \sqrt{4^2 - 2^2} \approx 3{,}46\,\mathrm{rad\,s}^{-1}$$

Since there is no steady-state response:

$$u_C(t) = u_C^n(t) = K\mathrm{e}^{-2t} \sin\left(3{,}46t + \varphi\right)\ \mathrm{V}$$

The values of the constants K and φ are calculated from the initial conditions:

$$u_C(0^+) = 2 = K \sin\varphi$$
$$u_C'(0^+) = 12 = -2K \sin\varphi + 3{,}46\, K \cos\varphi$$

The first equation is substituted in the second one, resulting in:

$$12 = -2 \cdot 2 + 3{,}46\, K \cos\varphi \;\; \rightarrow \;\; K \cos\varphi \approx 4{,}62$$

The value of φ is calculated consequently:

$$\frac{K \sin\varphi}{K \cos\varphi} = \tan\varphi = \frac{2}{4{,}62} \;\; \rightarrow \;\; \varphi \approx 23{,}41°$$

Once the value of φ is obtained, K is calculated:

$$K = \frac{2}{\sin\varphi} = \frac{2}{\sin(23{,}41°)} \approx 5{,}03\,\text{V}$$

So:

$$u_C(t) = 5{,}03 \cdot e^{-2t} \sin(3{,}46t + 23{,}41°)\ \text{V}$$

The time evolution of $u_C(t)$ for $R=1\,\Omega$ has been represented in Fig. 2.20.

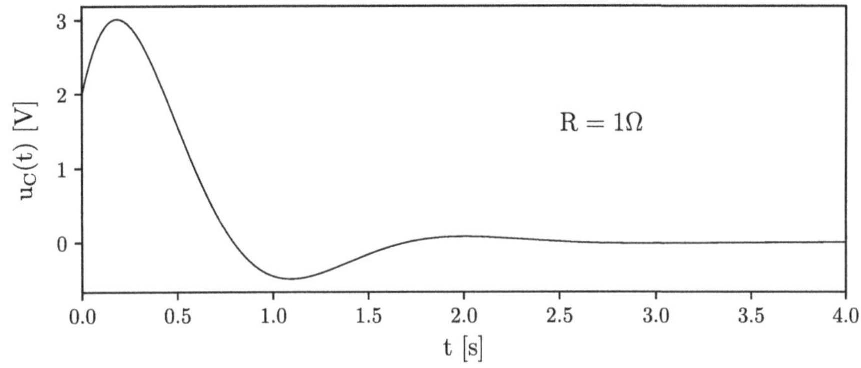

Fig. 2.20 Time evolution of $u_C(t)$ for $R=1\,\Omega$

(d) For $R=0\,\Omega$, the values of α and ω_0 are:

$$\alpha = \frac{R}{2L} = 0\,\text{s}^{-1} \ ; \ \omega_0 = \frac{1}{\sqrt{LC}} = \frac{1}{\sqrt{(1/4)\cdot(1/4)}} = 4\,\text{rad}\,\text{s}^{-1}$$

Since $\alpha=0$, the natural response is no damping:

$$u_C^n(t) = K\sin(\omega_0 t + \varphi)$$

Since there is no steady-state response:

$$u_C(t) = u_C^n(t) = K\sin(4t + \varphi)$$

The values of the constants K and φ are calculated from the initial conditions:

$$u_C(0^+) = 2 = K\sin\varphi$$
$$u'_C(0^+) = 12 = 4K\cos\varphi$$

To solve the system of equations, the quotient of both is calculated, obtaining the following value for φ:

$$\frac{K\sin\varphi}{4K\cos\varphi} = \frac{2}{12} \quad\rightarrow\quad \tan\varphi = \frac{2}{3} \quad\rightarrow\quad \varphi \approx 33{,}69^\circ$$

Once the value of φ is obtained, the value of K is consequently calculated:

$$K = \frac{2}{\sin\varphi} = \frac{2}{\sin(33{,}69^\circ)} \approx 3{,}61\,\mathrm{V}$$

So:

$$u_C(t) = 3{,}61\sin(4t + 33{,}69^\circ)\ \mathrm{V}$$

The time evolution of $u_C(t)$ for R=0 Ω has been represented in Fig. 2.21.

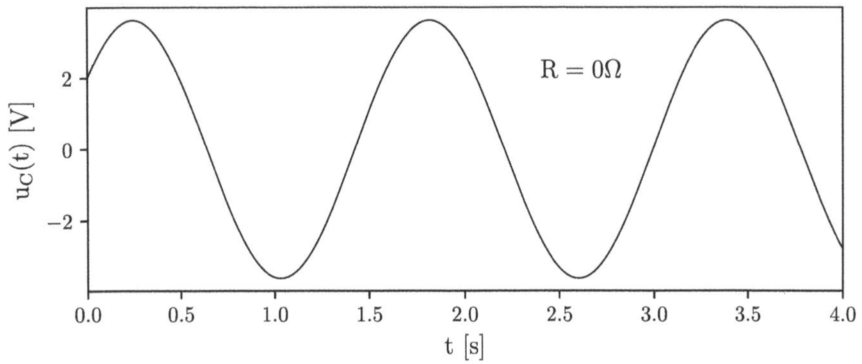

Fig. 2.21 Time evolution of $u_C(t)$ for R=0 Ω

2.4.3 Parallel RLC Without Excitation Sources

In the circuit represented in Fig. 2.22, given that $i_L(0^-)$=2 A and $u_C(0^-)$=400 V, calculate $i_L(t)$ for t>0 in the following cases:

(a) R=80 Ω
(b) R=100 Ω
(c) R=125 Ω
(d) R=∞ Ω

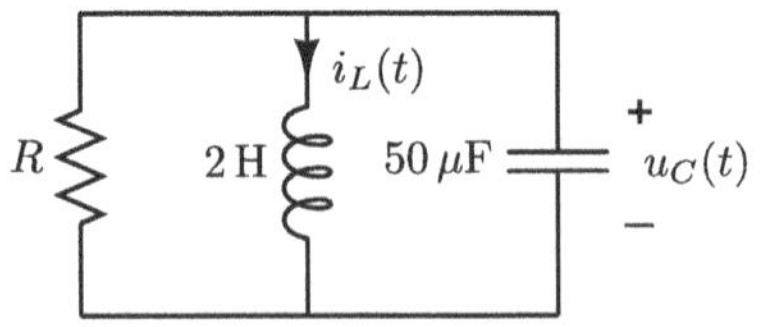

Fig. 2.22 Circuit considered in the problem under study

Solution. The response of a second order circuit is obtained as the sum of the steady-state response and the natural response. Since the circuit has no excitation sources the steady-state response is zero.

The natural response in a second order circuit is characterized by the value of the damping coefficient, α, and that of the resonant frequency, ω_0. Being a parallel RLC circuit, the values of these parameters are obtained based on the R, L and C according to:

$$\alpha = \frac{G}{2C} \ ; \ \omega_0 = \frac{1}{\sqrt{LC}}$$

where $G = 1/R$.

The initial conditions, $i_L(0^+)$ and $i'_L(0^+)$, are calculated by analyzing the circuit at t=0$^+$, which is obtained by substituting the capacitor for a voltage source of value $u_C(0^+)$ and the inductor for a current source of value $i_L(0^+)$. Except for an impulse-type response, the capacitor voltage and the inductor current do not change from t=0$^-$ to t=0$^+$. Thus:

$$i_L(0^+) = i_L(0^-) = 2\,\text{A} \ ; \ u_C(0^+) = u_C(0^-) = 400\,\text{V}$$

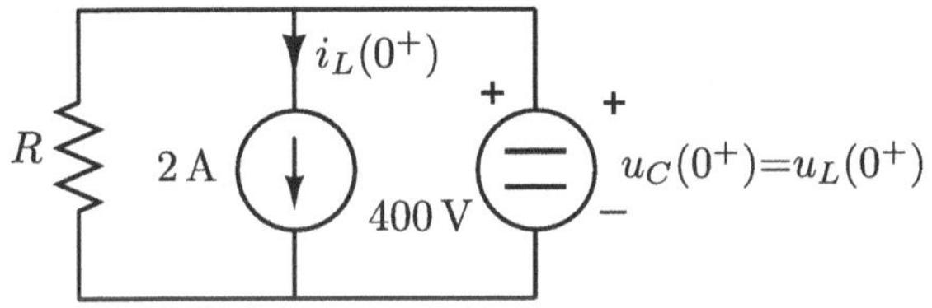

Fig. 2.23 Circuit at t=0$^+$

This way, the circuit at t=0$^+$ is that shown in Fig. 2.23.

Analyzing the circuit in Fig. 2.23 the voltage $u_L(0^+)$ is obtained and, consequently, the value of $i'_L(0^+)$:

$$i'_L(0^+) = \frac{u_L(0^+)}{L} = \frac{400}{2} = 200\,\text{A/s}$$

Since both $i_L(0^+)$ and $i'_L(0^+)$ are independent of R, their values will be valid for all the case studies.

(a) For R=80 Ω, the value of α and ω_0 are calculated as:

$$\alpha = \frac{G}{2C} = \frac{1/80}{2 \cdot 50 \cdot 10^{-6}} = 125\,\text{s}^{-1} \ ; \ \omega_0 = \frac{1}{\sqrt{LC}} = \frac{1}{2 \cdot 50 \cdot 10^{-6}} = 100\,\text{rad}\,\text{s}^{-1}$$

Since $\alpha > \omega_0$, the natural response is overdamped:

$$i_L^n(t) = K_1 \cdot e^{s_1 t} + K_2 \cdot e^{s_2 t}$$

where

$$s_1 = -\alpha + \sqrt{\alpha^2 - \omega_0^2} = -125 + \sqrt{125^2 - 100^2} = -200\,\mathrm{s}^{-1}$$
$$s_2 = -\alpha - \sqrt{\alpha^2 - \omega_0^2} = -125 - \sqrt{125^2 - 100^2} = -50\,\mathrm{s}^{-1}$$

Since there is no steady-state response:

$$i_L(t) = i_L^n(t) = K_1 \cdot e^{-200t} + K_2 \cdot e^{-50t}\ \mathrm{A}$$

The values of the constants K_1 and K_2 are calculated from the initial conditions:

$$\left.\begin{array}{l} i_L(0^+) = 2 = K_1 + K_2 \\ i_L'(0^+) = 200 = -200K_1 - 50K_2 \end{array}\right\} \Longrightarrow K_1 = -2\,\mathrm{A}\ ;\quad K_2 = 4\,\mathrm{A}$$

So:

$$i_L(t) = -2 \cdot e^{-200t} + 4 \cdot e^{-50t}\ \mathrm{A}$$

The time evolution of $i_L(t)$ for $R=80\,\Omega$ has been represented in Fig. 2.24.

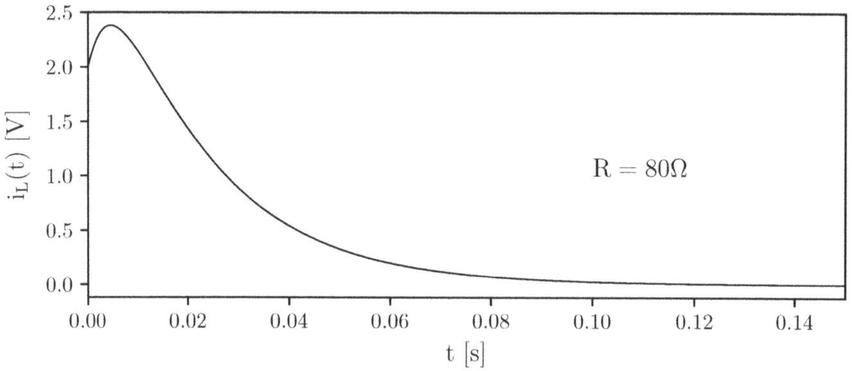

Fig. 2.24 Time evolution of $i_L(t)$ for $R=80\,\Omega$

(b) For $R=100\,\Omega$, the values of α and ω_0 are:

$$\alpha = \frac{G}{2C} = \frac{1/100}{2 \cdot 50 \cdot 10^{-6}} = 100\,\mathrm{s}^{-1}\ ;\quad \omega_0 = \frac{1}{\sqrt{LC}} = \frac{1}{2 \cdot 50 \cdot 10^{-6}} = 100\,\mathrm{rad\,s}^{-1}$$

Since $\alpha = \omega_0$, the natural response is critically damped:

$$i_L^n(t) = (K_1 + K_2 \cdot t) \cdot e^{-\alpha t}$$

Since there is no steady-state response:

$$i_L(t) = i_L^n(t) = (K_1 + K_2 \cdot t) \cdot e^{-100t} \text{ A}$$

The values of the constants K_1 and K_2 are calculated from the initial conditions:

$$\left.\begin{array}{l} i_L(0^+) = 2 = K_1 \\ i_L'(0^+) = 200 = -100K_1 + K_2 \end{array}\right\} \Longrightarrow K_1 = 2\,\text{A} \;;\; K_2 = 400\,\text{A/s}$$

So:

$$i_L(t) = (2 + 400 \cdot t) \cdot e^{-100t} \text{ A}$$

The time evolution of $i_L(t)$ for $R = 100\,\Omega$ has been represented in Fig. 2.25.

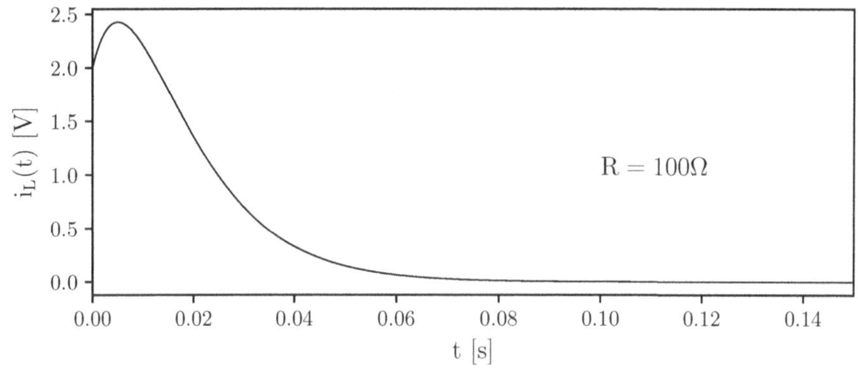

Fig. 2.25 Time evolution of $i_L(t)$ for $R = 100\,\Omega$

(c) For $R = 125\,\Omega$, the values of α and ω_0 are:

$$\alpha = \frac{G}{2C} = \frac{1/125}{2 \cdot 50 \cdot 10^{-6}} = 80\,\text{s}^{-1} \;;\; \omega_0 = \frac{1}{\sqrt{LC}} = \frac{1}{2 \cdot 50 \cdot 10^{-6}} = 100\,\text{rad}\,\text{s}^{-1}$$

Since $\alpha < \omega_0$, the natural response is underdamped:

$$i_L^n(t) = K e^{-\alpha t} \sin(\omega_d t + \varphi)$$

where:

$$\omega_d = \sqrt{\omega_0^2 - \alpha^2} = \sqrt{100^2 - 80^2} = 60\,\text{rad}\,\text{s}^{-1}$$

Since there is no steady-state response:

$$i_L(t) = i_L^n(t) = K\mathrm{e}^{-80t}\sin(60t + \varphi)\ \mathrm{A}$$

The values of the constants K and φ are calculated from the initial conditions:

$$i_L(0^+) = 2 = K\sin\varphi$$
$$i_L'(0^+) = 200 = -80K\sin\varphi + 60\,K\cos\varphi$$

The first equation is substituted in the second one, resulting in:

$$200 = -80\cdot 2 + 60\,K\cos\varphi \quad\rightarrow\quad K\cos\varphi = 6$$

The value of φ is calculated below:

$$\frac{K\sin\varphi}{K\cos\varphi} = \tan\varphi = \frac{2}{6} \quad\rightarrow\quad \varphi \approx 18{,}43^\circ$$

Once the value of φ is obtained, K is calculated as:

$$K = \frac{2}{\sin\varphi} = \frac{2}{\sin(18{,}43^\circ)} \approx 6{,}32\,\mathrm{A}$$

So:

$$i_L(t) = 6{,}32\cdot\mathrm{e}^{-80t}\sin(60t + 18{,}43^\circ)\ \mathrm{A}$$

The time evolution of $i_L(t)$ for R=125 Ω has been represented in Fig. 2.26.

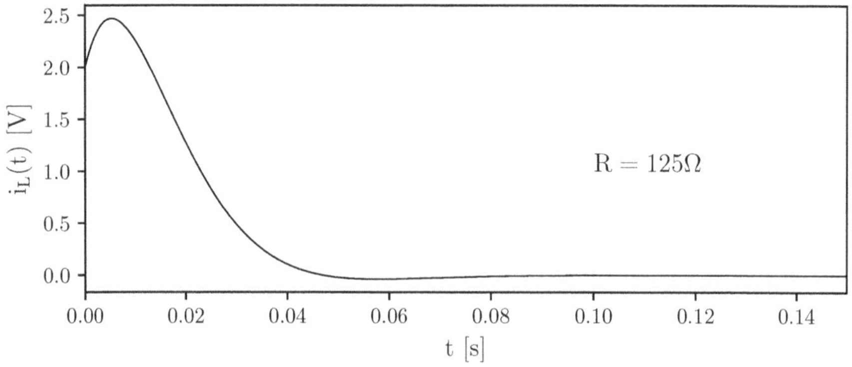

Fig. 2.26 Time evolution of $i_L(t)$ for R=125 Ω

(d) For $R=\infty$ Ω, the values of α and ω_0 are, in this case:

$$\alpha = \frac{G}{2C} = \frac{0}{2\cdot 50\cdot 10^{-6}} = 0\,\mathrm{s}^{-1}\ ;\ \omega_0 = \frac{1}{\sqrt{LC}} = \frac{1}{2\cdot 50\cdot 10^{-6}} = 100\,\mathrm{rad\,s}^{-1}$$

Since α=0, the natural response is no damping:

$$i_L^n(t) = K \sin(\omega_0 t + \varphi)$$

Since there is no steady-state response:

$$i_L(t) = i_L^n(t) = K \sin(100t + \varphi)$$

The values of the constants K and φ are calculated from the initial conditions:

$$i_L(0^+) = 2 = K \sin\varphi$$
$$i_L'(0^+) = 200 = 100\,K \cos\varphi$$

To solve the system of equations, the quotient of both is calculated, obtaining the value of φ:

$$\frac{K \sin\varphi}{100K \cos\varphi} = \frac{2}{200} \quad\rightarrow\quad \tan\varphi = 1 \quad\rightarrow\quad \varphi \approx 45^\circ$$

Once the value of φ is determined, K is calculated as:

$$K = \frac{2}{\sin\varphi} = \frac{2}{\sin(45^\circ)} \approx 2{,}83\,\text{A}$$

So:

$$i_L(t) = 2{,}83 \sin(100t + 45^\circ)\ \text{A}$$

The time evolution of $i_L(t)$ for R=∞ Ω has been represented in Fig. 2.27.

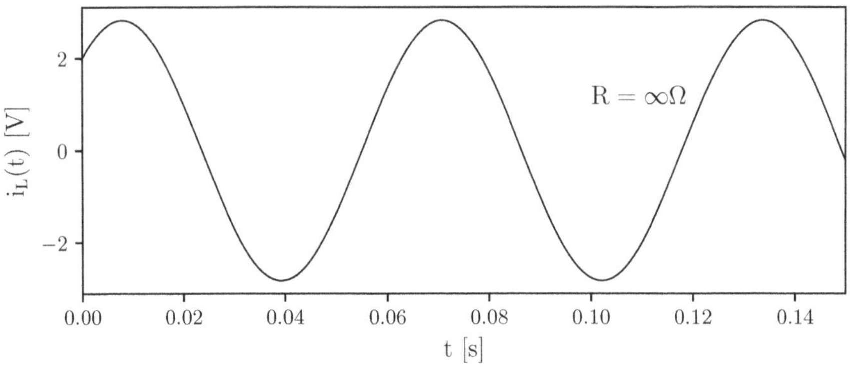

Fig. 2.27 Time evolution of $i_L(t)$ for R=∞ Ω

2.4.4 Overdamped Series RLC Under DC Supply

In the circuit represented in Fig. 2.28 calculate $i_L(t)$ and $u_C(t)$ for $t>0$, given that $u_C(0^-)=0\,\text{V}$.

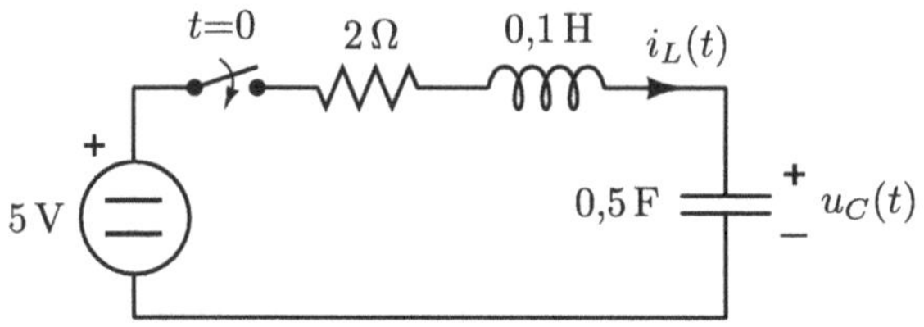

Fig. 2.28 Circuit considered in the problem under study

Solution. First, the initial conditions of the variables under study, $u_C(0^+)$, $u'_C(0^+)$, $i_L(0^+)$ and $i'_L(0^+)$, will be calculated. For this purpose, the circuit at $t=0^+$ is analyzed, which is obtained by substituting the capacitor for a voltage source of value $u_C(0^+)$ and the inductor for a current source of value $i_L(0^+)$. Except for an impulse-type response, the capacitor voltage and the current through the inductor do not change from $t=0^-$ to $t=0^+$. It can be noticed that, at $t=0$, the switch is open and consequently no current flows through the inductor. Therefore:

$$i_L(0^+) = i_L(0^-) = 0\,\text{A} \;\; ; \;\; u_C(0^+) = u_C(0^-) = 0\,\text{V}$$

This way, the circuit at $t=0^+$ is that shown in Fig. 2.29.

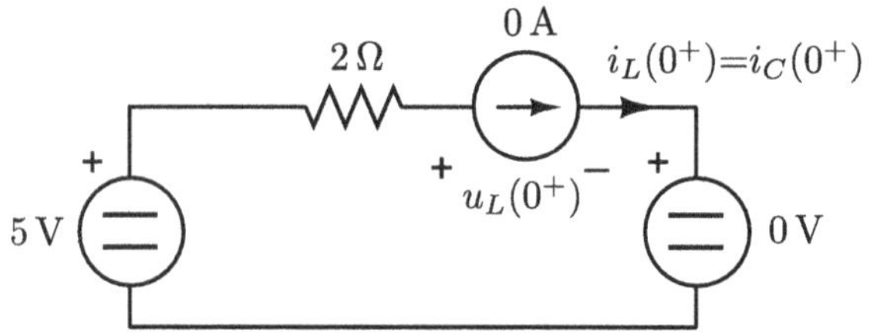

Fig. 2.29 Circuit at $t=0^+$

From the circuit in Fig. 2.29, the voltage $u_L(0^+)$ and the current $i_C(0^+)$ can be obtained and, consequently, the values of $i'_L(0^+)$ and $u'_C(0^+)$:

$$i'_L(0^+) = \frac{u_L(0^+)}{L} = \frac{5}{0{,}1} = 50\,\text{A/s}$$

$$u'_C(0^+) = \frac{i_C(0^+)}{C} = \frac{0}{0{,}5} = 0\,\text{V/s}$$

The steady-state variables are subsequently calculated. Since only a dc-type source of excitation is present, the circuit in steady-state dc regime will be solved, which is obtained by replacing the inductor with a short circuit and the capacitor with an open circuit, as shown in Fig. 2.30.

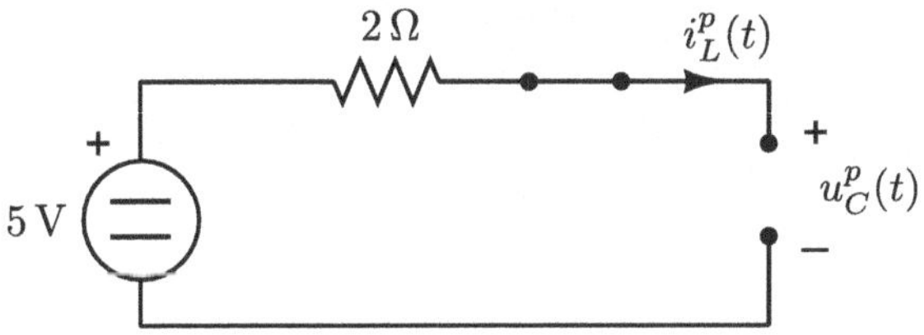

Fig. 2.30 Circuit in steady-state dc regime

From the circuit in Fig. 2.30:

$$i_L^p(t) = 0\,\text{A} \ ; \ u_C^p(t) = 5\,\text{V}$$

The natural response in a second order circuit is characterized by the value of the damping coefficient, α, and that of the resonant frequency, ω_0. Being a series RLC circuit, the values of these parameters are obtained based on R, L and C according to:

$$\alpha = \frac{R}{2L} = \frac{2}{2 \cdot 0{,}1} = 10\,\text{s}^{-1} \ ; \ \omega_0 = \frac{1}{\sqrt{LC}} = \frac{1}{\sqrt{0{,}1 \cdot 0{,}5}} = \sqrt{20} \approx 4{,}47\,\text{rad}\,\text{s}^{-1}$$

Since $\alpha > \omega_0$, the natural response is overdamped:

$$i_L^n(t) = K_{1L} \cdot \text{e}^{s_1 t} + K_{2L} \cdot \text{e}^{s_2 t}$$
$$u_C^n(t) = K_{1C} \cdot \text{e}^{s_1 t} + K_{2C} \cdot \text{e}^{s_2 t}$$

where

$$s_1 = -\alpha + \sqrt{\alpha^2 - \omega_0^2} = -10 + \sqrt{10^2 - (\sqrt{20})^2} \approx -1{,}05\,\text{s}^{-1}$$
$$s_2 = -\alpha - \sqrt{\alpha^2 - \omega_0^2} = -10 - \sqrt{10^2 - (\sqrt{20})^2} \approx -18{,}94\,\text{s}^{-1}$$

The complete response of the circuit is obtained as the sum of the steady-state response and the natural response, yielding:

$$i_L(t) = i_L^p(t) + i_L^n(t) = 0 + K_{1L} \cdot \text{e}^{-1{,}05t} + K_{2L} \cdot \text{e}^{-18{,}94t}$$
$$u_C(t) = u_C^p(t) + u_C^n(t) = 5 + K_{1C} \cdot \text{e}^{-1{,}05t} + K_{2C} \cdot \text{e}^{-18{,}94t}$$

The values of the constants K_{1L}, K_{2L}, K_{1C} and K_{2C} are calculated from the initial conditions:

$$\left.\begin{array}{l} i_L(0^+) = 0 = K_{1L} + K_{2L} \\ i'_L(0^+) = 50 = -1{,}05K_{1L} - 18{,}94K_{2L} \end{array}\right\} \Longrightarrow K_{1L} \approx 2{,}80\,\text{A} \;;\; K_{2L} \approx -2{,}80\,\text{A}$$

$$\left.\begin{array}{l} u_C(0^+) = 0 = 5 + K_{1C} + K_{2C} \\ u'_C(0^+) = 0 = -1{,}05K_{1C} - 18{,}94K_{2C} \end{array}\right\} \Longrightarrow K_{1C} \approx -5{,}29\,\text{V} \;;\; K_{2C} \approx 0{,}29\,\text{V}$$

Finally, for $t>0$:

$$i_L(t) = 2{,}8 \cdot \text{e}^{-1{,}05t} - 2{,}8 \cdot \text{e}^{-18{,}94t}\ \text{A}$$
$$u_C(t) = 5 - 5{,}29 \cdot \text{e}^{-1{,}05t} + 0{,}29 \cdot \text{e}^{-18{,}94t}\ \text{V}$$

The time evolution of $i_L(t)$ and $u_C(t)$ has been represented in Fig. 2.31.

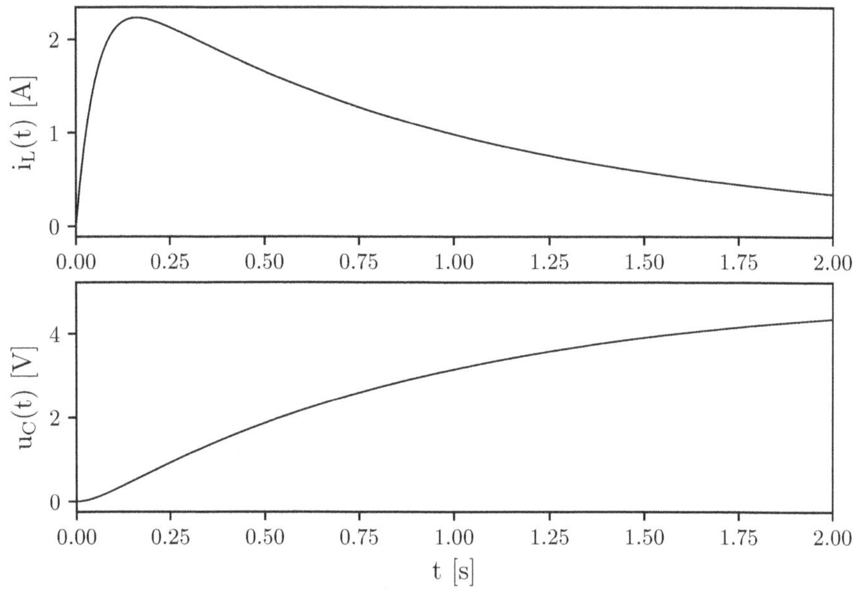

Fig. 2.31 Time evolution of $i_L(t)$ and $u_C(t)$

2.4.5 Underdamped Series RLC Under DC Supply

In the circuit represented in Fig. 2.32, with $i_L(0^-)$=9 A and $u_C(0^-)$=3 V, calculate $u_L(t)$ and $u_C(t)$ for $t>0$.

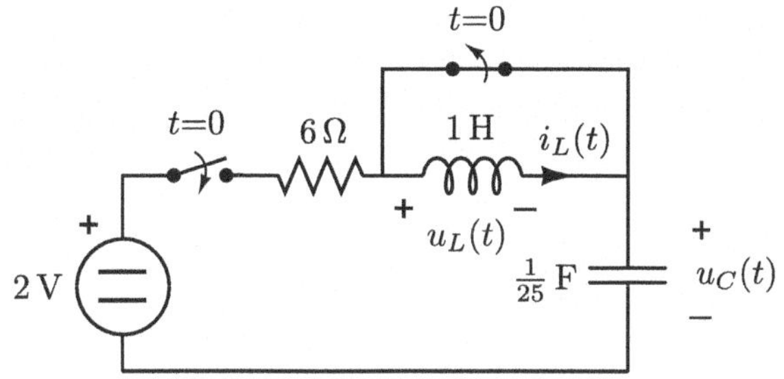

Fig. 2.32 Circuit considered in the problem under study

Solution. First, the initial conditions of the variables under study will be calculated: $u_C(0^+)$, $u'_C(0^+)$, $u_L(0^+)$ and $u'_L(0^+)$. For this purpose, the circuit at t=0^+ is analyzed, which is obtained by substituting the capacitor for a voltage source of value $u_C(0^+)$ and the inductor for a current source of value $i_L(0^+)$. Except for an impulse-type response, the capacitor voltage and the inductor current do not change from t=0^- to t=0^+. Therefore:

$$i_L(0^+) = i_L(0^-) = 9\,\text{A} \ ; \ \ u_C(0^+) = u_C(0^-) = 3\,\text{V}$$

This way, the circuit at t=0^+ is that shown in Fig. 2.33.

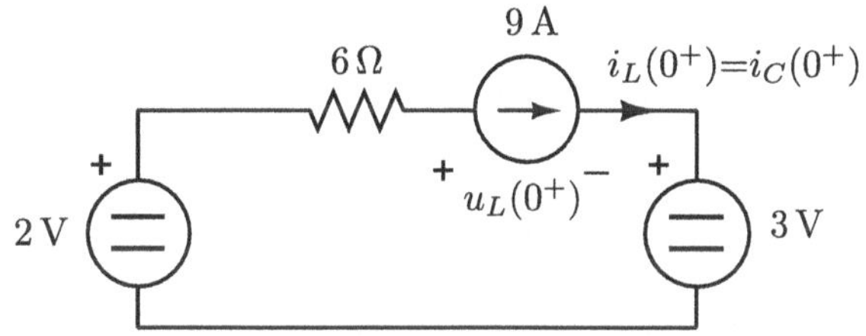

Fig. 2.33 Circuit at t=0^+

From the circuit of Fig. 2.33, $u_L(0^+)$ and $i_C(0^+)$ can be obtained and, consequently, the values of $i'_L(0^+)$ and $u'_C(0^+)$:

$$i'_L(0^+) = \frac{u_L(0^+)}{L} = \frac{2 - 6 \cdot 9 - 3}{1} = \frac{-55}{1} = -55\,\text{A/s}$$

$$u'_C(0^+) = \frac{i_C(0^+)}{C} = \frac{9}{1/25} = 225\,\text{V/s}$$

The derivative circuit at t=0^+ is subsequently analyzed, which is obtained by substituting the capacitor for a voltage source of $u'_C(0^+)$, the inductor for a current source of value $i'_L(0^+)$ and the source of excitation by a dc-source whose value is the derivative of its function at t=0^+. Since only a dc-type excitation source is present, its derivative is zero and therefore, it is replaced by a short circuit. This way, the derivative circuit at t=0^+ is that shown in Fig. 2.34.

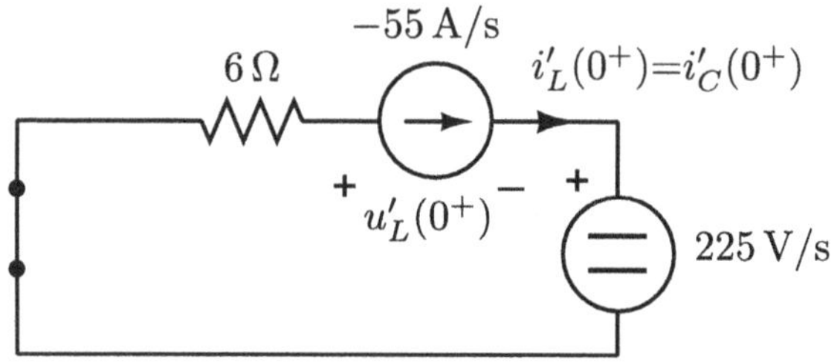

Fig. 2.34 Derivative circuit at t=0^+

From the circuit in Fig. 2.34, $u'_L(0^+)$ is obtained as:

$$u'_L(0^+) = -225 - 6 \cdot (-55) = 105 \text{ V/s}$$

The steady-state variables are subsequently calculated. Only a dc-type source of excitation is present, so that the dc circuit shown in Fig. 2.35 will be solved, where the inductor has been replaced by a short circuit and the capacitor by an open circuit.

From the circuit in Fig. 2.35, $u_L^p(t)$ and $u_C^p(t)$ are obtained as:

$$u_L^p(t) = 0\,\text{V} \;\; ; \;\; u_C^p(t) = 2\,\text{V}$$

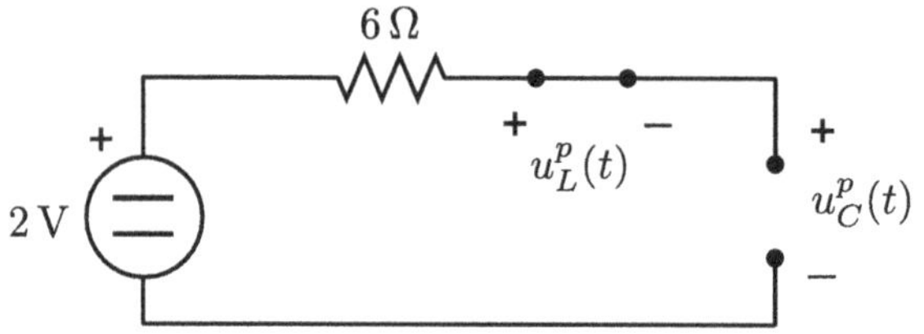

Fig. 2.35 Steady-state dc circuit

The natural response in a second order circuit is characterized by the value of the damping coefficient, α, and that of the resonant frequency, ω_0. Being a series RLC circuit, the values of these parameters are obtained based on R, L and C according to:

$$\alpha = \frac{R}{2L} = \frac{6}{2 \cdot 1} = 3\,\text{s}^{-1} \;\; ; \;\; \omega_0 = \frac{1}{\sqrt{LC}} = \frac{1}{\sqrt{1 \cdot 1/25}} = 5\,\text{rad}\,\text{s}^{-1}$$

Since $\alpha < \omega_0$, the natural response is underdamped:

$$u_L^n(t) = K_L e^{-\alpha t} \sin(\omega_d t + \varphi_L)$$
$$u_C^n(t) = K_C e^{-\alpha t} \sin(\omega_d t + \varphi_C)$$

where:

$$\omega_d = \sqrt{\omega_0^2 - \alpha^2} = \sqrt{5^2 - 3^2} = 4 \text{ rad s}^{-1}$$

The response of the circuit is obtained as the sum of the steady-state response and the natural response, yielding:

$$u_L(t) = u_L^p(t) + u_L^n(t) = 0 + K_L e^{-3t} \sin(4t + \varphi_L)$$
$$u_C(t) = u_C^p(t) + u_C^n(t) = 2 + K_C e^{-3t} \sin(4t + \varphi_C)$$

The values of the constants K_L, K_C, φ_L and φ_C are calculated from the initial conditions. Regarding K_L and φ_L:

$$u_L(0^+) = -55 = K_L \sin\varphi_L$$
$$u_L'(0^+) = 105 = -3K_L \sin\varphi_L + 4K_L \cos\varphi_L$$

The first equation is substituted in the second equation, resulting in:

$$105 = 165 + 4K_L \cos\varphi_L \quad \rightarrow \quad K_L \cos\varphi_L = -15$$

The value of φ_L is calculated below:

$$\frac{K_L \sin\varphi_L}{K_L \cos\varphi_L} = \tan\varphi_L = \frac{-55}{-15} \quad \rightarrow \quad \varphi_L \approx 74{,}74^\circ$$

Once the value of φ_L is obtained, K_L is calculated as:

$$K_L = \frac{-55}{\sin\varphi_L} = \frac{-55}{\sin(74{,}74^\circ)} \approx -57{,}01 \text{ V}$$

Regarding the constants K_C and φ_C:

$$u_C(0^+) = 3 = 2 + K_C \sin\varphi_C$$
$$u_C'(0^+) = 225 = -3K_C \sin\varphi_C + 4K_C \cos\varphi_C$$

Proceeding in the same way as in the previous case:

$$225 = -3 + 4K_C \cos\varphi_C \quad \rightarrow \quad K_C \cos\varphi_C = 57$$

Then the value of φ_C is calculated

$$\frac{K_C \sin \varphi_C}{K_C \cos \varphi_C} = \tan \varphi_C = \frac{1}{57} \quad \rightarrow \quad \varphi_C \approx 1^\circ$$

Once the value of φ_C is obtained, K_C is calculated as:

$$K_C = \frac{3-2}{\sin \varphi_C} = \frac{1}{\sin(1^\circ)} \approx 57{,}3\,\text{V}$$

So:

$$u_L(t) = -57{,}01 \cdot \mathrm{e}^{-3t} \sin\left(4t + 74{,}74^\circ\right)\,\text{V}$$
$$u_C(t) = 2 + 57{,}3 \cdot \mathrm{e}^{-3t} \sin\left(4t + 1^\circ\right)\,\text{V}$$

The time evolution of $u_L(t)$ and $u_C(t)$ has been represented in Fig. 2.36.

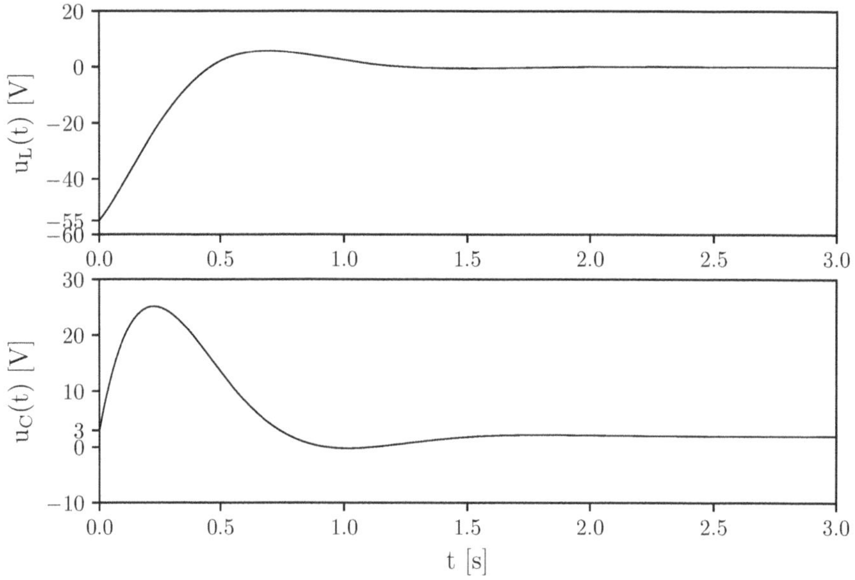

Fig. 2.36 Time evolution of $u_L(t)$ and $u_C(t)$

2.4.6 Overdamped Series RLC Under AC Supply

In the circuit represented in Fig. 2.37, calculate $i_L(t)$ for $t>0$, given that $e(t)=70\cos(40t)$ V, $i_L(0^-)=-2$ A and $u_C(0^-)=50$ V.

Solution. First, the initial conditions of the variable under study and its derivative will be calculated: $i_L(0^+)$ and $i'_L(0^+)$. For this purpose, the circuit at $t=0^+$ is analyzed,

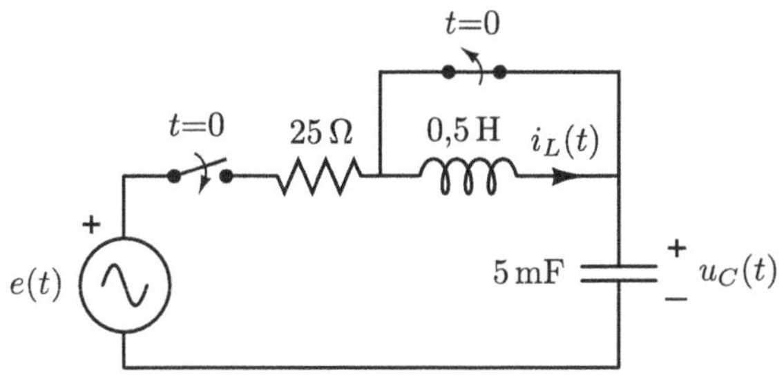

Fig. 2.37 Circuit considered in the problem under study

which is obtained by replacing the capacitor with a voltage source of value $u_C(0^+)$, the inductor with a current source of $i_L(0^+)$ and the ac voltage source by a dc source of value $e(0^+)$. Except for an impulse-type response, the capacitor voltage and the inductor current do not change from t=0^- to t=0^+. Therefore:

$$i_L(0^+) = i_L(0^-) = -2\,\text{A} \; ; \quad u_C(0^+) = u_C(0^-) = 50\,\text{V}$$

On the other hand

$$e(0^+) = 70\cos(40 \cdot 0) = 70\,\text{V}$$

This way, the circuit at t=0^+ is that shown in Fig. 2.38.

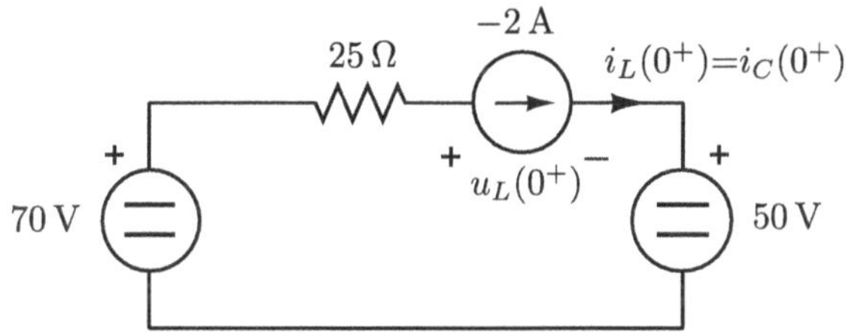

Fig. 2.38 Circuit at t=0^+

From the circuit in Fig. 2.38 the voltage $u_L(0^+)$ is calculated and, consequently, the value of $i'_L(0^+)$:

$$i'_L(0^+) = \frac{u_L(0^+)}{L} = \frac{70 + 2 \cdot 25 - 50}{0{,}5} = 140\,\text{A/s}$$

The steady-state current $i_L^p(t)$ is subsequently calculated. Since only an ac-type source of excitation is present, the circuit will be solved in steady-state ac regime, using phasor techniques. The circuit is obtained in the phasor domain, replacing the resistor, inductor and capacitor with their corresponding impedance

$$\overline{Z}_R = 25\,\Omega \;;\; \overline{Z}_L = j\omega L = j \cdot 40 \cdot 0{,}5 = 20j\,\Omega \;;\; \overline{Z}_C = \frac{-j}{\omega C} = \frac{-j}{40 \cdot 5 \cdot 10^{-3}} = -5j\,\Omega$$

and the voltage source by the following phasor:

$$\overline{E} = \frac{70}{\sqrt{2}}\angle 0^\circ \text{ V}$$

This way the circuit in the phasor domain is that shown in Fig. 2.39.

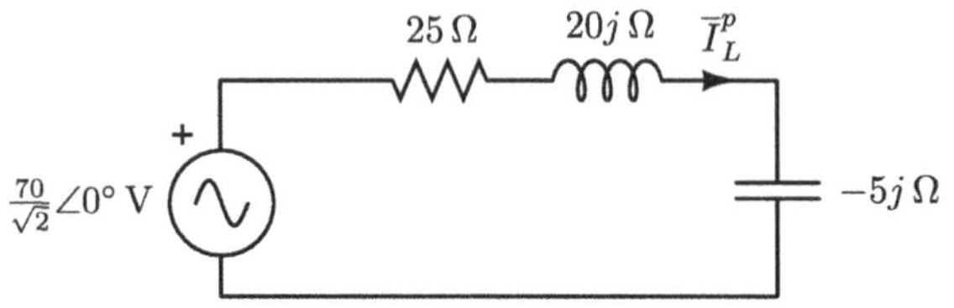

Fig. 2.39 Circuit in the phasor domain

From the circuit in Fig. 2.39, the current $\overline{I}_L^p$ can be obtained as:

$$\overline{I}_L^p = \frac{70/\sqrt{2}\angle 0^\circ}{25 + 20j - 5j} \approx 1{,}70\angle - 30{,}96^\circ \text{ A}$$

Once the variable of interest has been calculated in the phasor domain, the corresponding expression in the time domain is obtained:

$$i_L^p(t) = \sqrt{2} \cdot 1{,}7\cos(40t - 30{,}96^\circ) \text{ A}$$

The natural response in a second order circuit is characterized by the value of the damping coefficient, α, and that of the resonant frequency, ω_0. Being a series RLC circuit, the values of these parameters are obtained based on R, L and C according to:

$$\alpha = \frac{R}{2L} = \frac{25}{2 \cdot 0{,}5} = 25\,\text{s}^{-1} \;;\; \omega_0 = \frac{1}{\sqrt{LC}} = \frac{1}{\sqrt{0{,}5 \cdot 5 \cdot 10^{-3}}} = 20\,\text{rad}\,\text{s}^{-1}$$

Since $\alpha > \omega_0$, the natural response is overdamped:

$$i_L^n(t) = K_1 \cdot \text{e}^{s_1 t} + K_2 \cdot \text{e}^{s_2 t}$$

where

$$s_1 = -\alpha + \sqrt{\alpha^2 - \omega_0^2} = -25 + \sqrt{25^2 - 20^2} = -10\,\text{s}^{-1}$$

$$s_2 = -\alpha - \sqrt{\alpha^2 - \omega_0^2} = -25 - \sqrt{25^2 - 20^2} = -40\,\text{s}^{-1}$$

The response of the circuit is obtained as the sum of the steady-state response and the natural response, yielding:

$$i_L(t) = i_L^p(t) + i_L^n(t) = \sqrt{2} \cdot 1{,}7\cos(40t - 30{,}96°) + K_1 \cdot \mathrm{e}^{-10t} + K_2 \cdot \mathrm{e}^{-40t}$$

The values of the constants K_1 and K_2 are calculated from the initial conditions:

$$i_L(0^+) = -2 = \sqrt{2} \cdot 1{,}7\cos(-30{,}96°) + K_1 + K_2$$
$$i_L'(0^+) = 140 = -\sqrt{2} \cdot 1{,}7 \cdot 40\sin(-30{,}96°) - 10K_1 - 40K_2$$

Solving the system of equations:

$$K_1 \approx -2{,}39\,\mathrm{A} \;\; ; \;\; K_2 \approx -1{,}67\,\mathrm{A}$$

Finally, the complete response for $t>0$ is:

$$i_L(t) = \sqrt{2} \cdot 1{,}7\cos(40t - 30{,}96°) - 2{,}39 \cdot \mathrm{e}^{-10t} - 1{,}67 \cdot \mathrm{e}^{-40t}\,\mathrm{A}$$

The time evolution of $i_L(t)$ has been represented in Fig. 2.40.

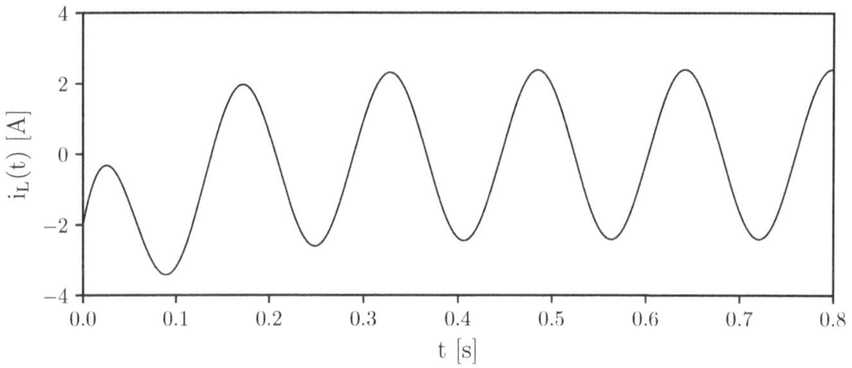

Fig. 2.40 Time evolution of $i_L(t)$

2.4.7 Critically Damped Series RLC Under DC Supply

The circuit depicted in Fig. 2.41 is in steady state when the switch is opened at t=0. Calculate the voltage $u_C(t)$ for t>0.

Solution. According to the problem statement, before the switch opens, the circuit is in steady state. Since, in this situation, only a dc-type source of excitation is present, the dc circuit shown in Fig. 2.42 will be solved in order to calculate the magnitudes

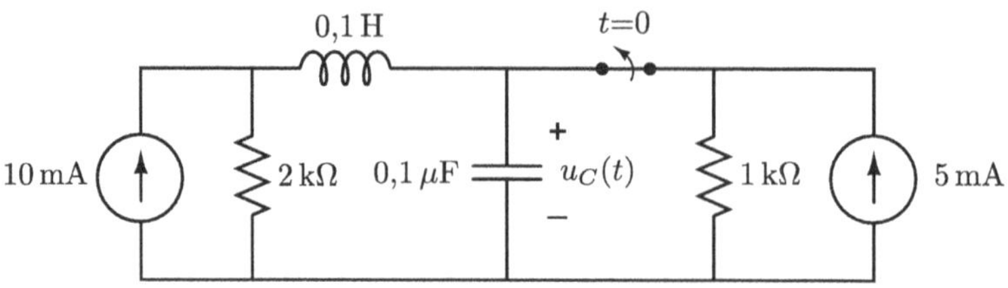

Fig. 2.41 Circuit considered in the problem under study

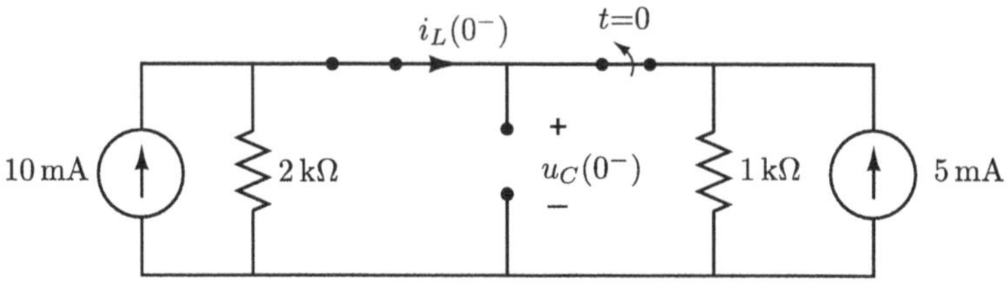

Fig. 2.42 Circuit in dc steady-state regime for t=0$^-$

of interest. The mentioned circuit has been obtained by replacing the capacitor with an open circuit and the inductor with a short circuit. It is important to remark that, besides the capacitor voltage, the inductor current has also been represented, given that its value will be required for the calculation of the initial conditions.

The circuit in Fig. 2.42 can be simplified by substituting both real current sources for their corresponding equivalent real voltage sources, as shown in Fig. 2.43, from this circuit:

$$i_L(0^-) = \frac{20 - 5}{2 \cdot 10^3 + 1 \cdot 10^3} = 5 \cdot 10^{-3}\,\text{A}$$

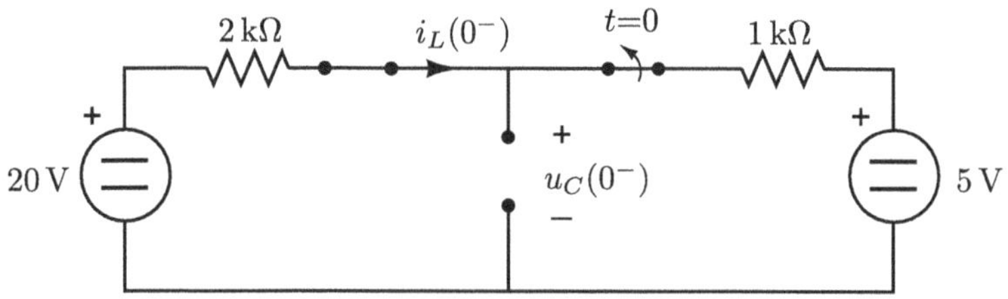

Fig. 2.43 Equivalent circuit in dc steady-state regime for t=0$^-$

Thus:

$$u_C(0^-) = 20 - 2 \cdot 10^3 \cdot i_L(0^-) = 20 - 2 \cdot 10^3 \cdot 5 \cdot 10^{-3} = 10\,\text{V}$$

Once the switch is opened ($t>0$), the resulting circuit is that shown in Fig. 2.44.

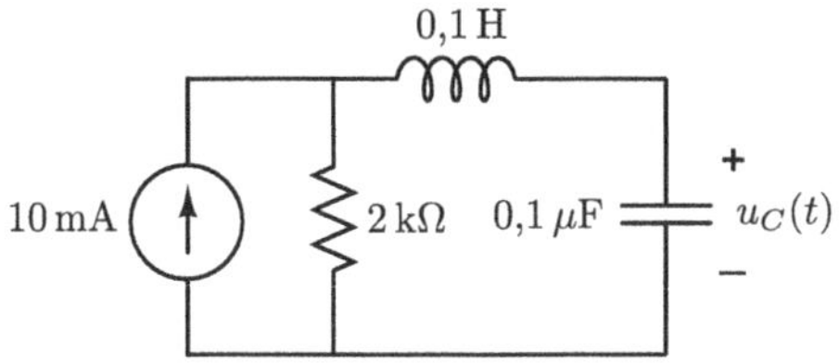

Fig. 2.44 Circuit for $t>0$

The initial conditions of $u_C(t)$ and its derivative, $u_C(0^+)$ and $u'_C(0^+)$, are subsequently calculated. For this purpose, the circuit at $t=0^+$ is analyzed, which is obtained by substituting the capacitor for a voltage source of value $u_C(0^+)$ and the inductor for a current source of value $i_L(0^+)$. Except for an impulse-type response, the capacitor voltage and the inductor current do not change from $t=0^-$ to $t=0^+$. Therefore:

$$i_L(0^+) = i_L(0^-) = 5 \cdot 10^{-3}\,\text{A} \;\; ; \;\; u_C(0^+) = u_C(0^-) = 10\,\text{V}$$

This way, the circuit at $t=0^+$ is that shown in Fig. 2.45.

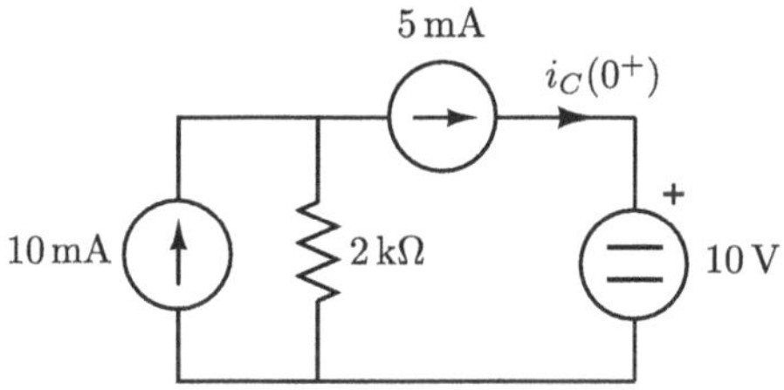

Fig. 2.45 Circuit at $t=0^+$

From the circuit in Fig. 2.45, the current $i_C(0^+)$ can be obtained and, consequently, the value of $u'_C(0^+)$:

$$u'_C(0^+) = \frac{i_C(0^+)}{C} = \frac{5 \cdot 10^{-3}}{0{,}1 \cdot 10^{-6}} = 5 \cdot 10^4\,\text{V/s}$$

Then, the steady-state value of the voltage $u_C(t)$ is calculated. Since only a dc-type source of excitation is present, the steady-state dc circuit shown in Fig. 2.46 will be solved, where the capacitor has been replaced by an open circuit and the inductor by a short circuit.

From the circuit in Fig. 2.46:

$$u_C^p(t) = 10 \cdot 10^{-3} \cdot 2 \cdot 10^3 = 20\,\text{V}$$

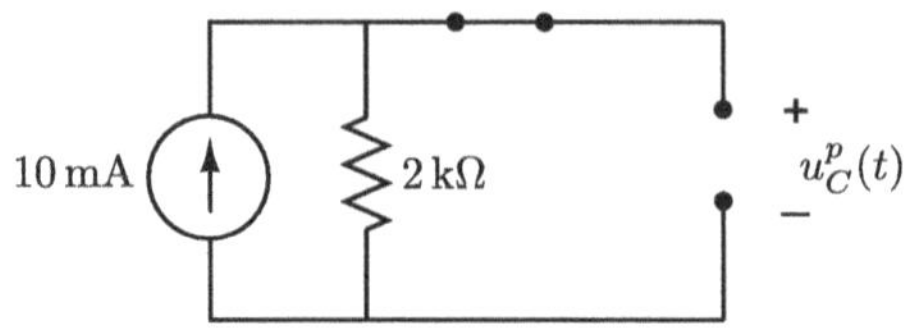

Fig. 2.46 Circuit in DC steady state

The natural response in a second order circuit is characterized by the value of the damping coefficient, α, and that of the resonant frequency, ω_0. Being a series RLC circuit, the values of these parameters are obtained based on R, L and C according to:

$$\alpha = \frac{R}{2L} = \frac{2 \cdot 10^3}{2 \cdot 0{,}1} = 10^4\,\text{s}^{-1} \;\; ; \;\; \omega_0 = \frac{1}{\sqrt{LC}} = \frac{1}{\sqrt{0{,}1 \cdot 0{,}1 \cdot 10^6}} = 10^4\,\text{rad}\,\text{s}^{-1}$$

Since $\alpha = \omega_0$, the natural response is critically damped:

$$u_C^n(t) = (K_1 + K_2 \cdot t) \cdot \mathrm{e}^{-\alpha t}$$

The response of the circuit is obtained as the sum of the steady-state response and the natural response, yielding:

$$u_C(t) = u_C^p(t) + u_C^n(t) = 20 + (K_1 + K_2 \cdot t) \cdot \mathrm{e}^{-10^4 t}$$

The values of the constants K_1 and K_2 are calculated from the initial conditions:

$$\left.\begin{aligned} u_C(0^+) &= 10 = 20 + K_1 \\ u_C'(0^+) &= 5 \cdot 10^4 = -10^4 K_1 + K_2 \end{aligned}\right\} \Longrightarrow K_1 = -10\,\text{V} \;\; ; \;\; K_2 = -5 \cdot 10^4\,\text{V/s}$$

So:

$$u_C(t) = 20 + (-10 - 5 \cdot 10^4 \cdot t) \cdot \mathrm{e}^{-10^4 t}\,\text{V}$$

The time evolution of $u_C(t)$ has been represented in Fig. 2.47.

2.4.8 Overdamped Parallel RLC Without Excitation Sources

The circuit represented in Fig. 2.48 is in steady state when, at t=0, the switches change their position. Calculate $u_C(0^+)$, $i_C(0^+)$, $i_L(0^+)$, $i_R(0^+)$ and $u_C(10\,\mu\text{s})$.

Solution. According to the problem statement, before the switches change their position, the circuit is in steady state. Since, in this situation, only a dc-type source of excitation is present, the steady state dc circuit shown in Fig. 2.49 will be solved, where the capacitor has been replaced by an open circuit and the inductor by a short circuit.

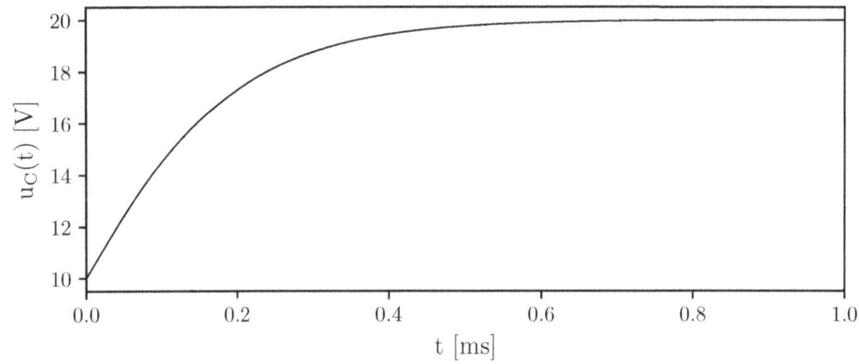

Fig. 2.47 Time evolution of $u_C(t)$

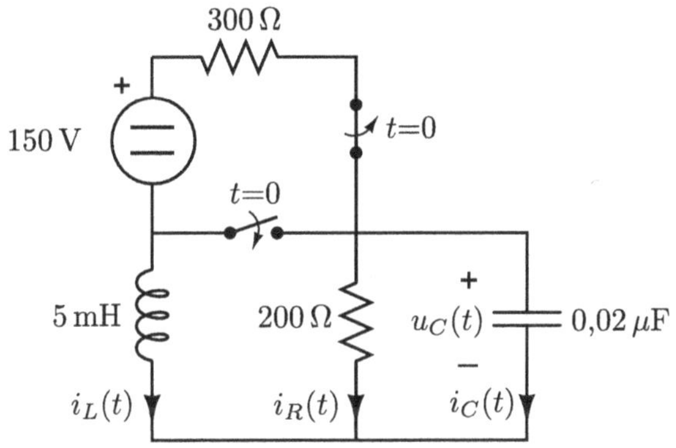

Fig. 2.48 Circuit considered in the problem under study

From the circuit in Fig. 2.49:

$$i_L(0^-) = \frac{-150}{300+200} = -0{,}3\ \text{A}$$

$$u_C(0^-) = 150 \cdot \frac{200}{300+200} = 60\ \text{V}$$

Once the switches have changed their position, $t>0$, the resulting circuit is that shown in Fig. 2.50.

To calculate the value of the variables at $t=0^+$, the circuit at that instant will be analyzed, which is obtained by substituting the capacitor for a voltage source of value $u_C(0^+)$ and the inductor by a current source of value $i_L(0^+)$. Except for an impulse-type response, the capacitor voltage and the inductor current do not change from $t=0^-$ to $t=0^+$. Therefore:

$$i_L(0^+) = i_L(0^-) = -0{,}3\ \text{A} \;\; ; \;\; u_C(0^+) = u_C(0^-) = 60\ \text{V}$$

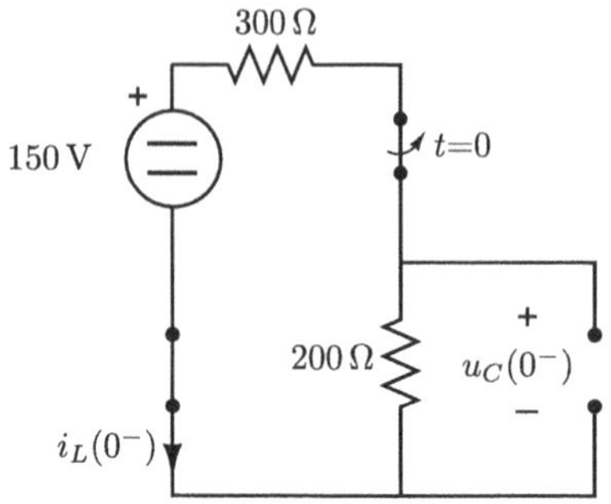

Fig. 2.49 Circuit in steady state dc regime at t=0$^-$

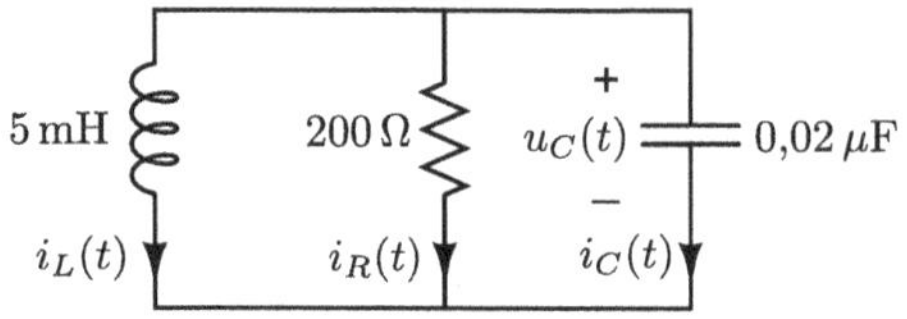

Fig. 2.50 Circuit for t>0

This way, the circuit at t=0$^+$ is that shown in Fig. 2.51, from which the following results are obtained:

$$i_R(0^+) = \frac{60}{200} = 0{,}3\,\text{A}$$
$$i_C(0^+) = -(-0{,}3 + i_R(0^+)) = -(-0{,}3 + 0{,}3) = 0\,\text{A}$$

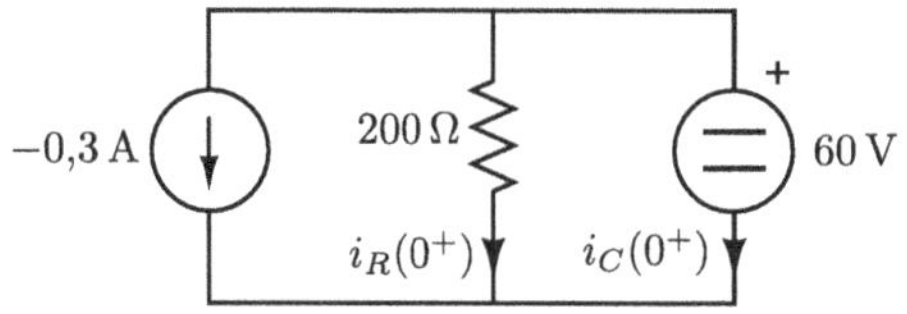

Fig. 2.51 Circuit at t=0$^+$

Using $i_C(0^+)$, the value of $u'_C(0^+)$ can be calculated as:

$$u'_C(0^+) = \frac{i_C(0^+)}{C} = \frac{0}{0{,}02 \cdot 10^{-6}} = 0\,\text{V/s}$$

The natural response in a second order circuit is characterized by the value of the damping coefficient, α, and that of the resonant frequency, ω_0. Being a parallel RLC circuit, the values of these parameters are obtained based on R, L and C according to:

$$\alpha = \frac{G}{2C} = \frac{1/200}{2 \cdot 0{,}02 \cdot 10^{-6}} = 125 \cdot 10^3\,\mathrm{s}^{-1}$$

$$\omega_0 = \frac{1}{\sqrt{LC}} = \frac{1}{\sqrt{5 \cdot 10^{-3} \cdot 0{,}02 \cdot 10^{-6}}} = 100 \cdot 10^3\,\mathrm{rad\,s}^{-1}$$

where $G=1/R$.

Since $\alpha>\omega_0$, the natural response is overdamped:

$$u_C^n(t) = K_1 \cdot \mathrm{e}^{s_1 t} + K_2 \cdot \mathrm{e}^{s_2 t}$$

where

$$s_1 = -\alpha + \sqrt{\alpha^2 - \omega_0^2} = -125 \cdot 10^3 + \sqrt{(125 \cdot 10^3)^2 - (100 \cdot 10^3)^2} = -5 \cdot 10^4\,\mathrm{s}^{-1}$$

$$s_2 = -\alpha - \sqrt{\alpha^2 - \omega_0^2} = -125 \cdot 10^3 - \sqrt{(125 \cdot 10^3)^2 - (100 \cdot 10^3)^2} = -2 \cdot 10^5\,\mathrm{s}^{-1}$$

Since the circuit for $t>0$ has no excitation sources, the steady-state response is null, yielding:

$$u_C(t) = u_C^n(t) = K_1 \cdot \mathrm{e}^{-5 \cdot 10^4 t} + K_2 \cdot \mathrm{e}^{-2 \cdot 10^5 t}\,\mathrm{V}$$

The values of the constants K_1 and K_2 are calculated from the initial conditions:

$$\left.\begin{aligned} u_C(0^+) &= 60 = K_1 + K_2 \\ u_C'(0^+) &= 0 = -5 \cdot 10^4 K_1 - 2 \cdot 10^5 K_2 \end{aligned}\right\} \Longrightarrow K_1 = 80\,\mathrm{V}\ ;\ K_2 = -20\,\mathrm{V}$$

Therefore, for $t>0$:

$$u_C(t) = 80 \cdot \mathrm{e}^{-5 \cdot 10^4 t} - 20 \cdot \mathrm{e}^{-2 \cdot 10^5 t}\,\mathrm{V}$$

Finally, the complete response of $u_C(t)$ for $t=10\,\mu$s is:

$$u_C(10\,\mu\mathrm{s}) = 80 \cdot \mathrm{e}^{-5 \cdot 10^4 \cdot 10 \cdot 10^{-6}} - 20 \cdot \mathrm{e}^{-2 \cdot 10^5 \cdot 10 \cdot 10^{-6}} \approx 45{,}8\,\mathrm{V}$$

The time evolution of $u_C(t)$ has been represented in Fig. 2.52.

2.4.9 Critically Damped Parallel RLC Under AC Supply

The circuit represented in Fig. 2.53 is in steady state when, at $t=0$, the switches change their position. Given that $e(t)=326\sin(100t - 30°)$ V, calculate the current $i_C(t)$ for $t>0$.

According to the problem statement, before the switches change their position, the circuit is in steady state. Since, in this situation, only a dc-type source of excitation

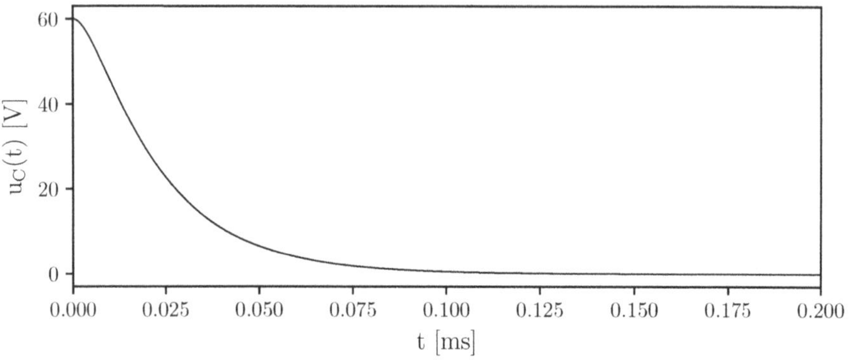

Fig. 2.52 Time evolution of $u_C(t)$

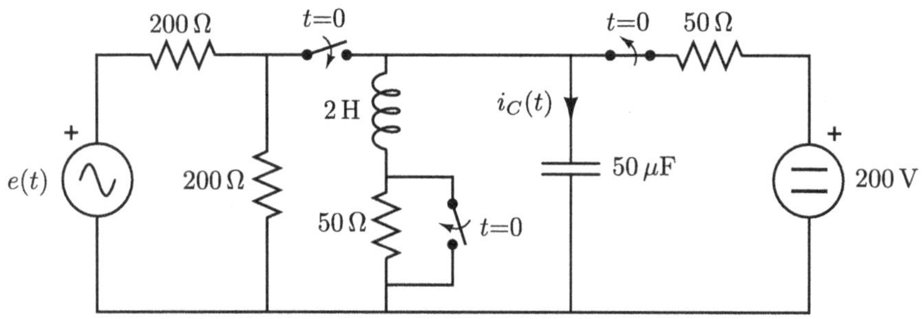

Fig. 2.53 Circuit considered in the problem under study

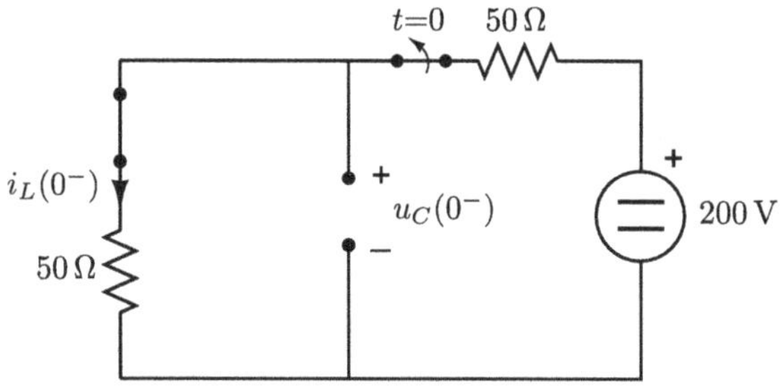

Fig. 2.54 Circuit in steady state dc regime at t=0$^-$

is present, the dc circuit shown in Fig. 2.54 will be solved, where the capacitor has been replaced by an open circuit and the inductor by a short circuit.

It should be remarked that, at t=0$^-$, it is interesting to calculate the inductor current and the capacitor voltage, since these values will be used to solve the circuit in t=0$^+$.

From the circuit in Fig. 2.54:

$$i_L(0^-) = \frac{200}{50+50} = 2\,\text{A}$$
$$u_C(0^-) = 50 \cdot i_L(0^-) = 50 \cdot 2 = 100\,\text{V}$$

Once the switches change their position ($t>0$), the resulting circuit is that shown in Fig. 2.55.

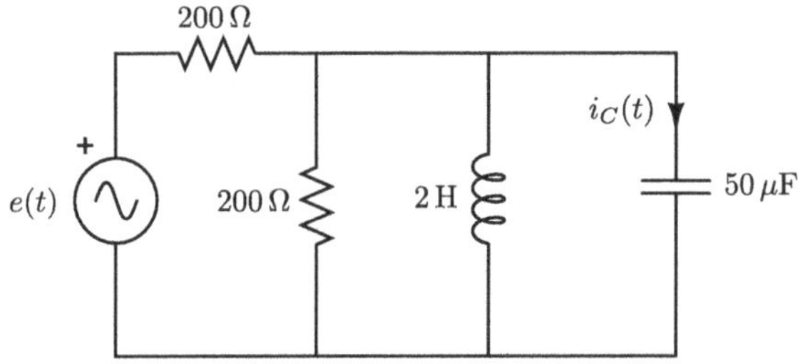

Fig. 2.55 Circuit for $t>0$

The initial conditions for $i_C(t)$, that is $i_C(0^+)$ and $i'_C(0^+)$, are subsequently calculated. For this purpose, the circuit at $t=0^+$ is analyzed, which is obtained by substituting the capacitor for a voltage source of value $u_C(0^+)$, the inductor for a current source of value $i_L(0^+)$ and the ac voltage source by a dc source whose value is $e(0^+)$. Except for an impulse-type response, the capacitor voltage and the inductor current do not change from $t=0^-$ to $t=0^+$. Therefore:

$$i_L(0^+) = i_L(0^-) = 2\,\text{A} \;\; ; \;\; u_C(0^+) = u_C(0^-) = 100\,\text{V}$$

This way, the circuit at $t=0^+$ is that shown in Fig. 2.56.

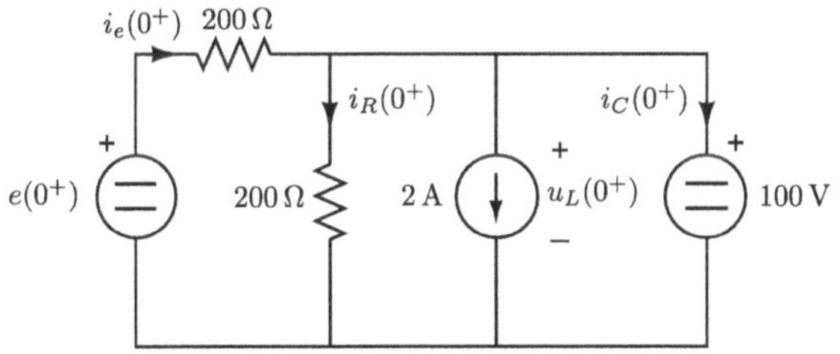

Fig. 2.56 Circuit at $t=0^+$

From the circuit in Fig. 2.56:

$$i_R(0^+) = \frac{100}{200} = 0{,}5\,\text{A}$$
$$i_e(0^+) = \frac{326\sin(-30°) - 100}{200} = \frac{-163-100}{200} = -1{,}315\,\text{A}$$

So:

$$i_C(0^+) = i_e(0^+) - i_R(0^+) - 2 = -1{,}315 - 0{,}5 - 2 = -3{,}815\,\text{A}$$

On the other hand, from the circuit in Fig. 2.56:

$$u_L(0^+) = 100\,\text{V}$$

Using $i_C(0^+)$ and $u_L(0^+)$, the values of $i'_L(0^+)$ and $u'_C(0^+)$ are obtained as:

$$i'_L(0^+) = \frac{u_L(0^+)}{L} = \frac{100}{2} = 50\,\text{A/s}$$
$$u'_C(0^+) = \frac{i_C(0^+)}{C} = \frac{-3{,}815}{50 \cdot 10^{-6}} = -76\,300\,\text{V/s}$$

Then, the derivative circuit at t=0^+ is analyzed, which is obtained by substituting the capacitor for a voltage source of value $u'_C(0^+)$, the inductor for a current source of value $i'_L(0^+)$ and the ac source of excitation by a dc source whose value is the derivative of its own function at t=0^+. With these premises, the derivative circuit at t=0^+ is that shown in Fig. 2.57.

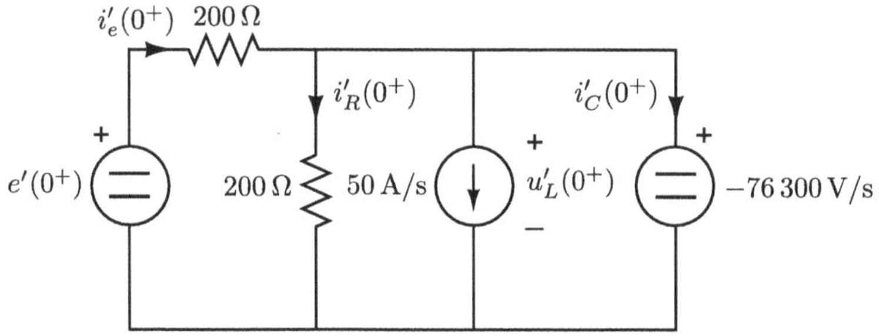

Fig. 2.57 Derivative circuit at t=0^+

From the circuit in Fig. 2.56:

$$i'_R(0^+) = \frac{-76\,300}{200} = -381{,}5\,\text{A/s}$$
$$i'_e(0^+) = \frac{326 \cdot 100\cos(-30°) + 76\,300}{200} \approx \frac{28\,232{,}43 + 76\,300}{200} = 522{,}66\,\text{A/s}$$

So:

$$i'_C(0^+) = i'_e(0^+) - i'_R(0^+) - 50 = 522{,}66 + 381{,}5 - 50 = 854{,}16\,\text{A/s}$$

The steady-state current $i_C^p(t)$ is subsequently calculated. Since only an ac-type source of excitation is present, the resulting circuit will be solved in steady-state ac regime, using the circuit in the phasor domain shown in Fig. 2.58.

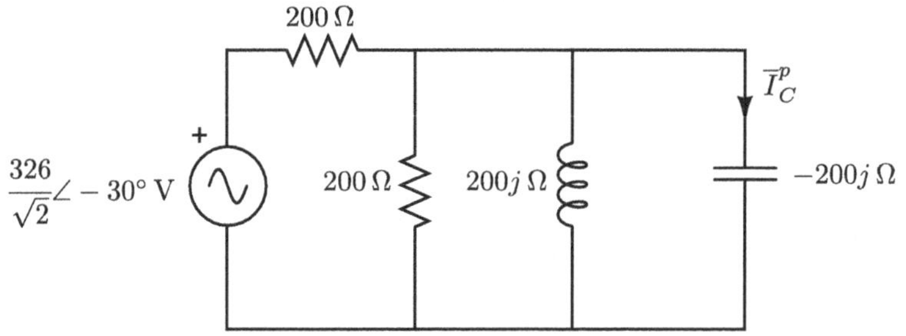

Fig. 2.58 Circuit in the phasor domain

The circuit depicted in Fig. 2.58 can be simplified by substituting the real voltage source for its corresponding equivalent real current source, as shown in Fig. 2.59.

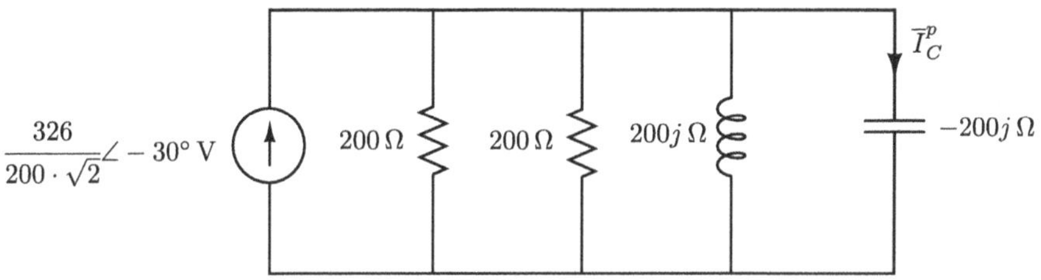

Fig. 2.59 Simplified circuit in the phasor domain

From the circuit in Fig. 2.58, the current $\overline{I}_C^p$ can be obtained using the concept of current divider:

$$\overline{I}_C^p = \frac{326}{200\cdot\sqrt{2}}\angle-30° \cdot \frac{(1/-200j)}{(1/100)+(1/200j)+(1/-200j)} \approx 0{,}58\angle60° \text{ A}$$

Once the variable under study in the phasor domain has been calculated, the corresponding expression in the time domain is obtained:

$$i_C^p(t) = \sqrt{2}\cdot 0{,}58\sin(100t + 60°) \text{ A}$$

The natural response in a second order circuit is characterized by the value of the damping coefficient, α, and that of the resonant frequency, ω_0. Being a parallel RLC circuit, the values of these parameters are obtained based on R, L and C according to:

$$\alpha = \frac{G_{eq}}{2C} = \frac{(1/200) + (1/200)}{2 \cdot 50 \cdot 10^{-6}} = 100\,\mathrm{s}^{-1}$$
$$\omega_0 = \frac{1}{\sqrt{LC}} = \frac{1}{\sqrt{2 \cdot 50 \cdot 10^{-6}}} = 100\,\mathrm{rad\,s}^{-1}$$

where $G_{eq}{=}1/R_{eq}$.

Since $\alpha{=}\omega_0$, the natural response is critically damped:

$$i_C^n(t) = (K_1 + K_2 \cdot t) \cdot \mathrm{e}^{-\alpha t}$$

The response of the circuit is obtained as the sum of the steady-state response and the natural response, yielding:

$$i_C(t) = i_C^p(t) + i_C^n(t) = \sqrt{2} \cdot 0{,}58 \sin(100t + 60°) + (K_1 + K_2 \cdot t) \cdot \mathrm{e}^{-100t}$$

The values of the constants K_1 and K_2 are calculated from the initial conditions:

$$\left.\begin{array}{l} i_C(0^+) = -3{,}815 = \sqrt{2} \cdot 0{,}58 \sin(60°) + K_1 \\ i_C'(0^+) = 854{,}16 = \sqrt{2} \cdot 0{,}58 \cdot 100 \cos(60°) - 100K_1 + K_2 \end{array}\right\} \Longrightarrow \begin{array}{l} K_1 \approx -4{,}5\,\mathrm{A} \\ K_2 \approx 361\,\mathrm{A/s} \end{array}$$

So:

$$i_C(t) = \sqrt{2} \cdot 0{,}58 \sin(100t + 60°) + (-4{,}5 + 361 \cdot t) \cdot \mathrm{e}^{-100t}\,\mathrm{A}$$

The time evolution of $i_C(t)$ has been represented in Fig. 2.60.

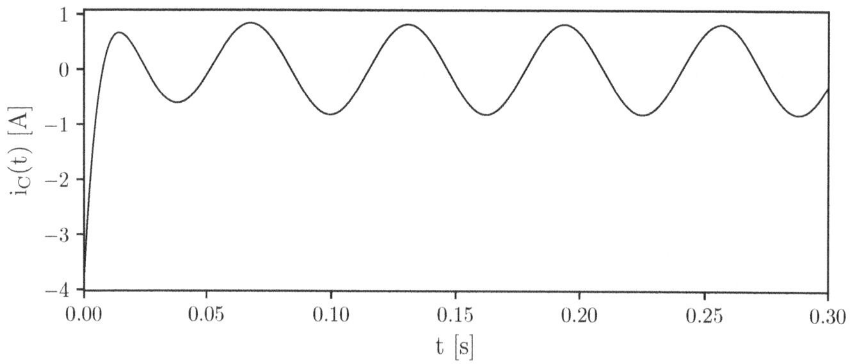

Fig. 2.60 Time evolution of $i_C(t)$

2.4.10 Overdamped Series RLC with Thévenin DC Equivalent

In the circuit represented in Fig. 2.61, calculate the capacitor voltage, $u_C(t)$, for $t{>}0$, given that $u_C(0^-){=}20\,\mathrm{V}$.

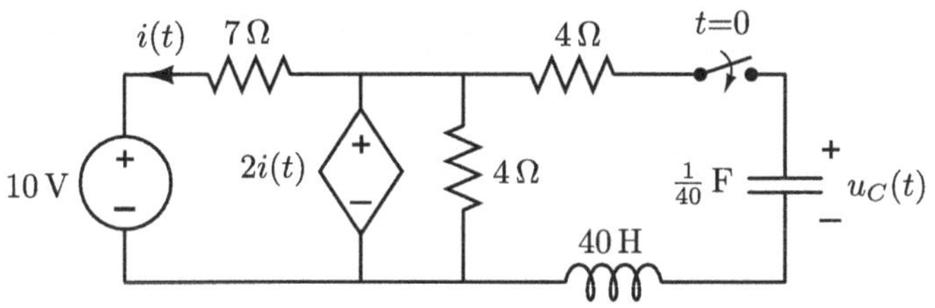

Fig. 2.61 Circuit considered in the problem under study

Solution. To simplify the study of the transient, the Thévenin equivalent seen from the terminals of the inductor-capacitor assembly will be obtained, as shown in Fig. 2.62.

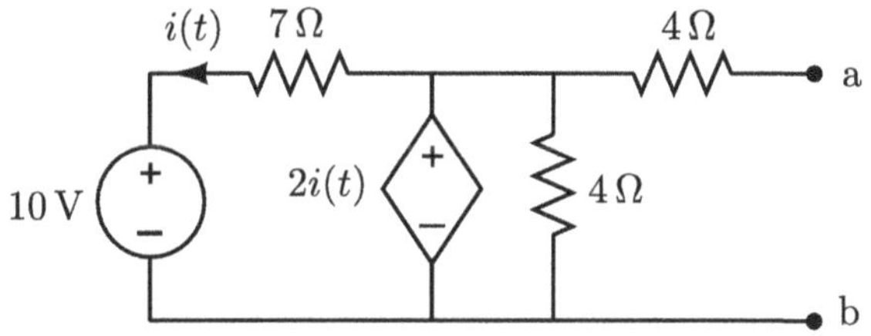

Fig. 2.62 Circuit seen from the terminals of the inductor-capacitor assembly

In order to obtain the Thévenin equivalent, the open-circuit voltage will be calculated and, consequently, the short-circuit current. Regarding the calculation of the open-circuit voltage between terminals a and b, the circuit to be solved is that shown in Fig. 2.63.

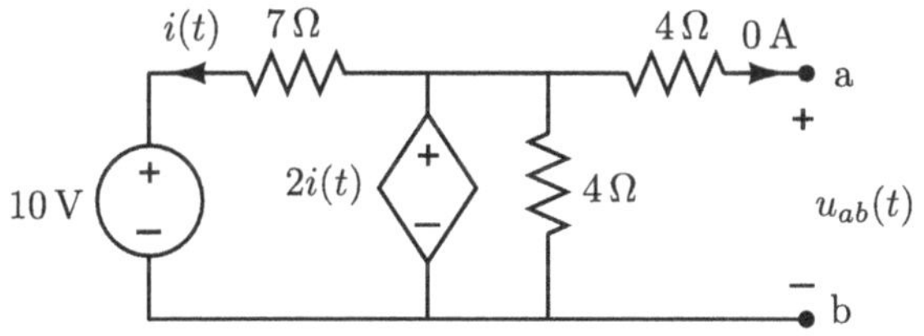

Fig. 2.63 Circuit used for the calculation of the open-circuit voltage between terminals a and b

From the circuit in Fig. 2.63, Kirchhoff's voltage law can be applied:

$$10 = -7i(t) + 2i(t) \;\Rightarrow\; i(t) = \frac{-10}{5} = -2\,\text{A}$$

On the other hand:

$$u_{ab}(t) = 2i(t) = 2 \cdot (-2) = -4\,\text{V}$$

Then, the short-circuit current between terminals a and b will be calculated, as shown in Fig. 2.64.

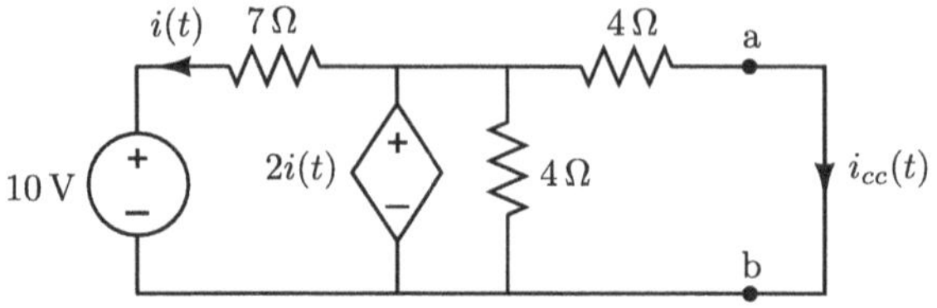

Fig. 2.64 Circuit used for the calculation of short-circuit current between terminals a and b

From the circuit in Fig. 2.64, Kirchhoff's voltage law can be applied:

$$10 = -7i(t) + 2i(t) \;\Rightarrow\; i(t) = \frac{-10}{5} = -2\,\mathrm{A}$$

On the other hand:

$$2i(t) = 4 \cdot i_{cc}(t) \;\Rightarrow\; i_{cc}(t) = \frac{2i(t)}{4} = \frac{2 \cdot (-2)}{4} = -1\,\mathrm{A}$$

Once the open-circuit voltage and the short-circuit current between terminals a and b have been obtained, the equivalent resistance (Thévenin resistance) is calculated as follows:

$$R_{eq} = \frac{u_{ab}(t)}{i_{cc}(t)} = \frac{-4}{-1} = 4\,\Omega$$

This way, the series RLC circuit in Fig. 2.65 is obtained.

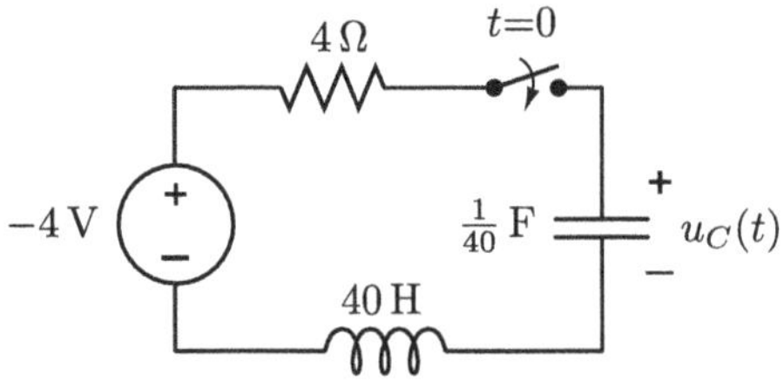

Fig. 2.65 Resulting RLC circuit

Then, the initial conditions of the variable under study, $u_C(0^+)$ and $u'_C(0^+)$, will be calculated. For this purpose, the circuit at t=0^+ is analyzed, which is obtained by substituting the capacitor for a voltage source of value $u_C(0^+)$ and the inductor for a current source of value $i_L(0^+)$. Except for an impulse-type response, the capacitor voltage and the inductor current do not change from t=0^- to t=0^+. Therefore:

$$i_L(0^+) = i_L(0^-) = 0\,\mathrm{A}\ ;\ \ u_C(0^+) = u_C(0^-) = 20\,\mathrm{V}$$

This way, the circuit at t=0$^+$ is that shown in Fig. 2.66.

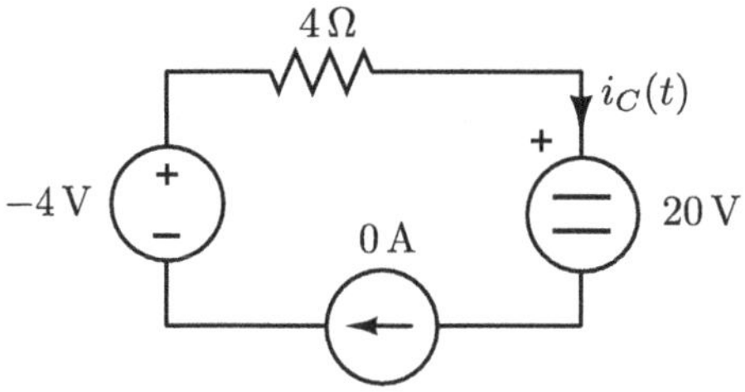

Fig. 2.66 Circuit at t=0$^+$

From the circuit in Fig. 2.66, the current $i_C(0^+)$ is obtained and, consequently, the value of $u'_C(0^+)$ is calculated as:

$$u'_C(0^+) = \frac{i_C(0^+)}{C} = \frac{0}{1/40} = 0\,\text{V/s}$$

Then, the steady state-value of the voltage $u_C(t)$ is calculated. Since only a dc-type source of excitation is present, the dc circuit shown in Fig. 2.67 will be solved, where the inductor has been replaced by a short circuit and the capacitor by an open circuit.

From the circuit in Fig. 2.67:

$$u^p_C(t) = -4\,\text{V}$$

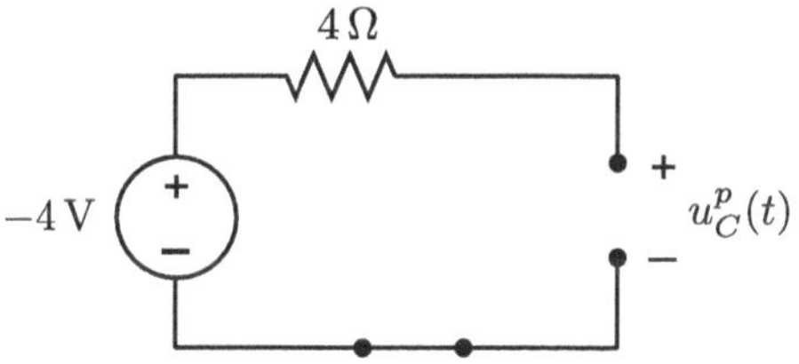

Fig. 2.67 Circuit in dc steady-state regime

The natural response in a second order circuit is characterized by the value of the damping coefficient, α, and that of the resonant frequency, ω_0. Being a series RLC circuit, the values of these parameters are obtained based on R, L and C according to:

$$\alpha = \frac{R}{2L} = \frac{4}{2 \cdot 40} = 0{,}05\,\text{s}^{-1} \quad ; \quad \omega_0 = \frac{1}{\sqrt{LC}} = \frac{1}{\sqrt{40 \cdot 1/40}} = 1\,\text{rad}\,\text{s}^{-1}$$

Since $\alpha < \omega_0$, the natural response is underdamped:

$$u^n_C(t) = K\text{e}^{-\alpha t} \sin{(\omega_d t + \varphi)}$$

where:

$$\omega_d = \sqrt{\omega_0^2 - \alpha^2} = \sqrt{1^2 - 0{,}05^2} \approx 1\,\text{rad}\,\text{s}^{-1}$$

The complete response of the circuit is obtained as the sum of the steady-state response and the natural response, yielding:

$$u_C(t) = u_C^p(t) + u_C^n(t) = -4 + K\text{e}^{-0{,}05t}\sin(t+\varphi)$$

The values of the constants K and φ are calculated from the initial conditions:

$$u_C(0^+) = 20 = -4 + K\sin\varphi$$
$$u'_C(0^+) = 0 = -0{,}05K\sin\varphi + K\cos\varphi$$

To solve the system of equations, the term $K\sin\varphi$ in the first equation is substituted in the second one, resulting in:

$$0 = -0{,}05\cdot 24 + K\cos\varphi \quad \rightarrow \quad K\cos\varphi = 1{,}2$$

The value of φ is calculated below:

$$\frac{K\sin\varphi}{K\cos\varphi} = \tan\varphi = \frac{24}{1{,}2} \quad \rightarrow \quad \varphi \approx 87{,}14^\circ$$

Once the value of φ is obtained, K is calculated as:

$$K = \frac{24}{\sin\varphi} = \frac{24}{\sin(87{,}14^\circ)} \approx 24{,}03\,\text{V}$$

So:

$$u_C(t) = -4 + 24{,}03\cdot\text{e}^{-0{,}05t}\sin(t + 87{,}13^\circ)\ \text{V}$$

The time evolution of $u_C(t)$ has been graphically represented in Fig. 2.68.

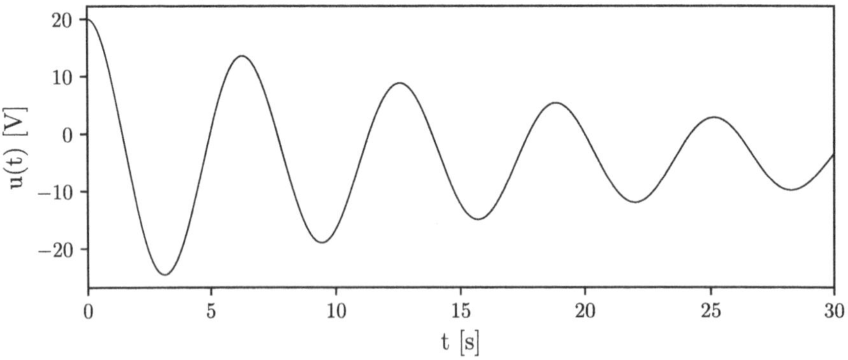

Fig. 2.68 Time evolution of $u_C(t)$

2.4.11 RLC Without Damping and Without Excitation Sources

In the circuit represented in Fig. 2.69 the switch closes at t=0. Given that $u_{C1}(0^-)$=1 V and $u_{C2}(0^-)$=4 V, calculate $i_L(t)$ for t>0.

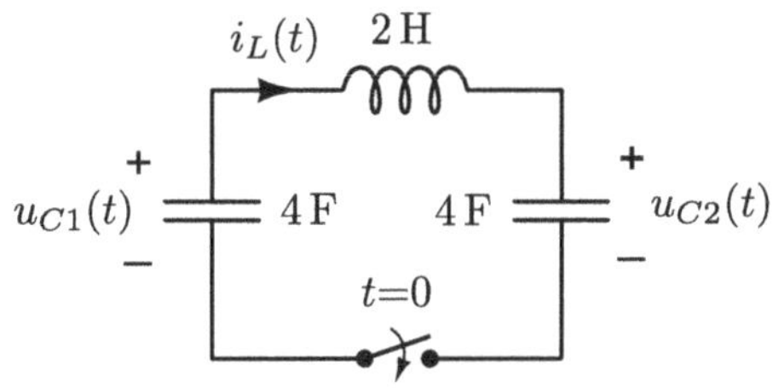

Fig. 2.69 Circuit considered in the problem under study

Solution. According to the problem statement, before the switch closes, the capacitors are charged to a certain initial voltage. Except for an impulse-type response, the voltages of both capacitor do not change when the switch is closed, therefore:

$$u_{C1}(0^+) = u_{C1}(0^-) = 1\,\text{V}$$
$$u_{C2}(0^+) = u_{C2}(0^-) = 4\,\text{V}$$

On the other hand, the voltage of each capacitor can be expressed as follows

$$u_{C1}(t) = u_{C1}(0^+) + \frac{1}{C_1}\int_{0^+}^{t} i_{C1}(\lambda)d\lambda = 1 + u^d_{C1}(t)$$
$$u_{C2}(t) = u_{C2}(0^+) + \frac{1}{C_2}\int_{0^+}^{t} i_{C2}(\lambda)d\lambda = 4 + u^d_{C2}(t)$$

where $u^d_{C1}(t)$ and $u^d_{C2}(t)$ are the voltages of the two discharged capacitors. This result leads to the circuit in Fig. 2.70.

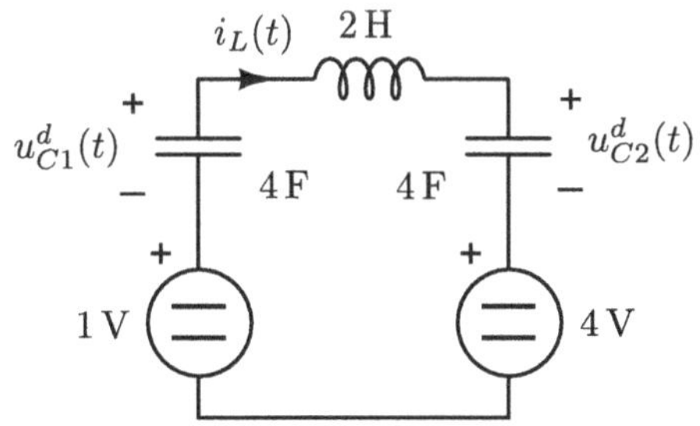

Fig. 2.70 Circuit for $t > 0$

In the circuit depicted in Fig. 2.70, the discharged capacitors and the voltage sources corresponding to the initial conditions can be associated in series, resulting in the circuit represented in Fig. 2.71, where

$$u^d_{Ceq}(t) = u^d_{C1}(t) + u^d_{C1}(t)$$

and

$$C_{eq} = \frac{C_1 \cdot C_2}{C_1 + C_2} = \frac{4 \cdot 4}{4+4} = 2\,\mathrm{F}$$

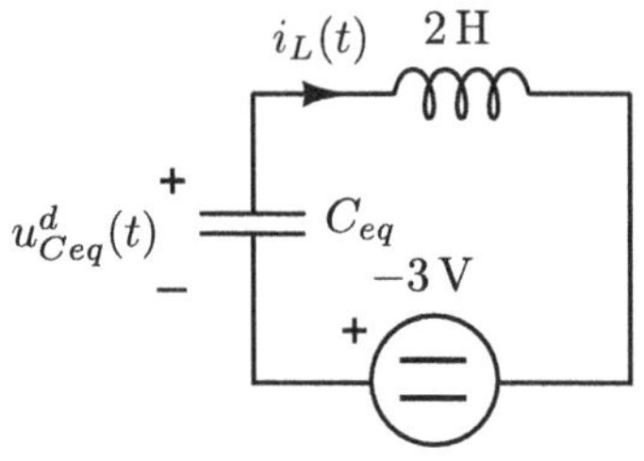

Fig. 2.71 Resulting circuit for t>0

Then, the initial conditions of the variable under study, $i_L(0^+)$ and $i'_L(0^+)$, will be calculated. For this purpose, the circuit at t=0$^+$ is analyzed, which is obtained by substituting the capacitor for a voltage source of value $u^d_{C_eq}(0^+)$ and the inductor for a current source of value $i_L(0^+)$. Except for an impulse-type response, the inductor current does not change from t=0$^-$ to t=0$^+$. Therefore:

$$i_L(0^+) = i_L(0^-) = 0\,\mathrm{A}$$

On the other hand, $u^d_{C_{eq}}(0^+)$ corresponds to the voltage of the equivalent discharged capacitor at time t=0$^+$. So that

$$u^d_{Ceq}(0^+) = 0\,\mathrm{V}$$

This way, the circuit at t=0$^+$ is that shown in Fig. 2.72.

From the circuit in Fig. 2.72, the voltage $u_L(0^+)$ is calculated and, consequently, the value of $i'_L(0^+)$ can be obtained as:

$$i'_L(0^+) = \frac{u_L(0^+)}{L} = \frac{-3}{2} = -1{,}5\,\mathrm{A/s}$$

The steady-state current, $i^p_L(t)$, is obtained by solving the dc steady-state circuit shown in Fig. 2.73, where the inductor has been replaced by a short circuit and the capacitor by a open circuit.

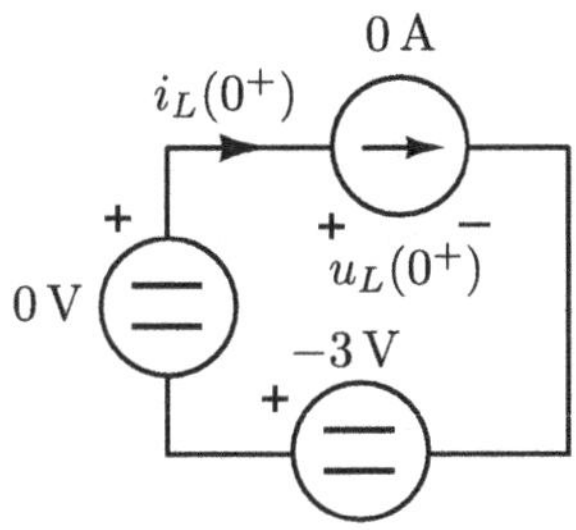

Fig. 2.72 Circuit at t=0^+

From the circuit in Fig. 2.73:

$$i_L^p(t) = 0\,\text{A}$$

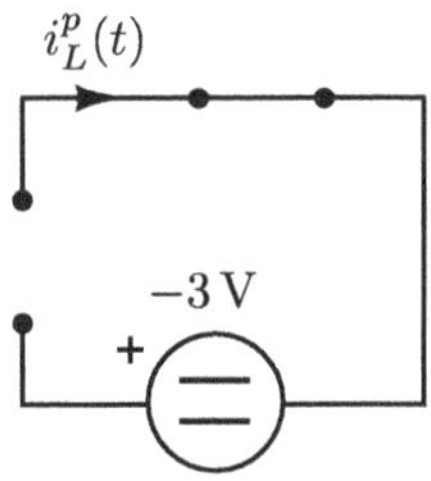

Fig. 2.73 Circuit in dc steady-state regime

The natural response in a second order circuit is characterized by the value of the damping coefficient, α, and that of the resonant frequency, ω_0. In this case, it can be verified that the circuit lacks resistance, so that:

$$\alpha = 0\,\text{s}^{-1}$$

On the other hand, the resonant frequency is calculated as follows:

$$\omega_0 = \frac{1}{\sqrt{LC_{eq}}} = \frac{1}{\sqrt{2 \cdot 2}} = 0{,}5\,\text{rad}\,\text{s}^{-1}$$

Since α=0, the natural response is no damping:

$$i_L^n(t) = K \sin(\omega_0 t + \varphi)$$

The complete response of the circuit is obtained as the sum of the steady-state response and the natural response, yielding:

$$i_L(t) = i_L^p(t) + i_L^n(t) = K \sin(0{,}5t + \varphi)$$

The values of the constants K and φ are calculated from the initial conditions:

$$i_L(0^+) = 0 = K \sin\varphi$$
$$i'_L(0^+) = -1{,}5 = 0{,}5K \cos\varphi$$

Solving the previous system of equations, the values of K and φ are obtained:

$$\varphi = 0^\circ \;;\; K = \frac{-1{,}5}{0{,}5} = -3\,\text{A}$$

So:

$$i_L(t) = -3 \sin(0{,}5t)\ \text{A}$$

The time evolution of $i_L(t)$ has been graphically represented in Fig. 2.74.

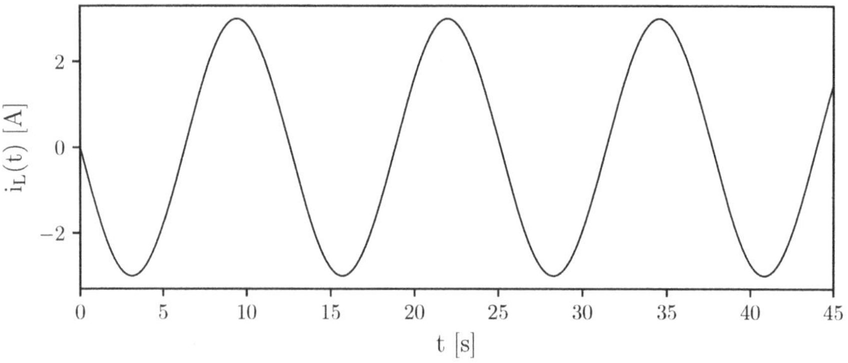

Fig. 2.74 Time evolution of $i_L(t)$

2.4.12 Overdamped Series RLC Under AC Supply

The circuit represented in Fig. 2.75 is in steady state when, at t=2 s, the switch opens. Calculate $u_C(t)$ for t>2 s given that $u_g(t)=\sqrt{2}\cdot 100\sin(t)$ V.

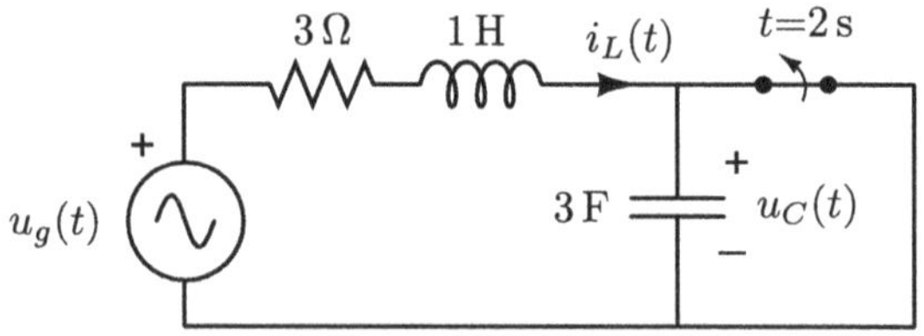

Fig. 2.75 Circuit considered in the problem under study

Solution. First, the inductor current and the capacitor voltage will be obtained before the switch opens (t=2 s$^-$). For the capacitor, it can be verified that its voltage is zero, therefore:

$$u_C(2\,\mathrm{s}^-) = 0\,\mathrm{V}$$

Regarding the inductor, it will be considered that the circuit is in steady state before the switch opens. Since, in this situation, only an ac-type source of excitation source is present, the corresponding circuit in ac steady-state regime will be solved in the phasor domain, as shown in Fig. 2.76.

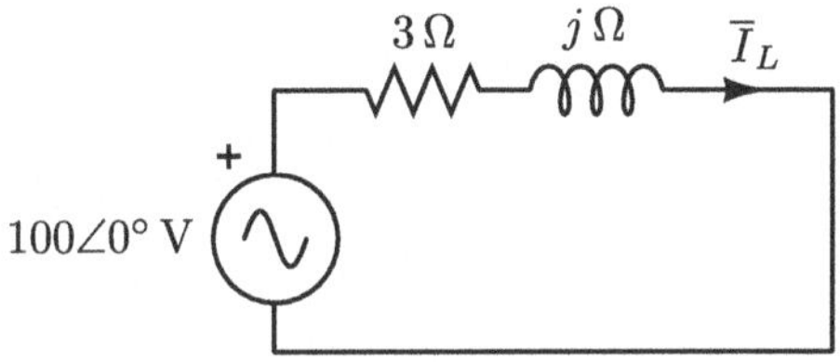

Fig. 2.76 Circuit in the phasor domain

From the circuit in Fig. 2.76:

$$\overline{I}_L = \frac{100\angle 0^\circ}{3+j} \approx 31{,}62\angle -18{,}43^\circ\ \mathrm{A}$$

Once the current in the phasor domain has been calculated, the corresponding expression in the time domain is obtained

$$i_L(t) = 31{,}62\sqrt{2}\sin(t - 18{,}43^\circ)\ \mathrm{A}$$

whose value at the instant t=2 s$^-$ is:

$$i_L(2\,\mathrm{s}^-) = 31{,}62\sqrt{2}\sin\left(\frac{2\cdot 360^\circ}{2\pi} - 18{,}43^\circ\right) \approx 44{,}46\,\mathrm{A}$$

It is important to remark that, in the previous operation, the radians have been converted to degrees.

Once the switch is opened (t>2), the resulting circuit is that shown in Fig. 2.77.

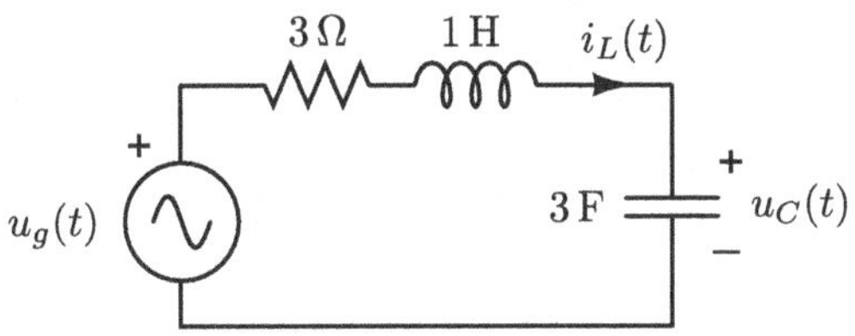

Fig. 2.77 Circuit for t>2

Then, the initial conditions of the variable under study and its derivative, $u_C(2\,s^+)$ and $u'_C(2\,s^+)$, will be calculated. For this purpose, the circuit at $t{=}2\,s^+$ is analyzed, which is obtained by substituting the capacitor for a voltage source of value $u_C(2\,s^+)$, the inductor by a current source of value $i_L(2\,s^+)$ and the ac voltage source by a dc voltage source whose value is $u_g(2\,s^+)$. Except for an impulse-type response, the capacitor voltage and the inductor current do not change from $t{=}2\,s^-$ to $t{=}2\,s^+$. Therefore:

$$i_L(2\,s^+) = i_L(2\,s^-) = 44{,}46\,\text{A} \;\; ; \;\; u_C(2\,s^+) = u_C(2\,s^-) = 0\,\text{V}$$

On the other hand:

$$u_g(2\,s^+) = \sqrt{2}\cdot 100\sin\left(\frac{2\cdot 360^\circ}{2\pi}\right) \approx 128{,}59\,\text{V}$$

This way, the circuit at $t{=}2\,s^+$ is that shown in Fig. 2.78.

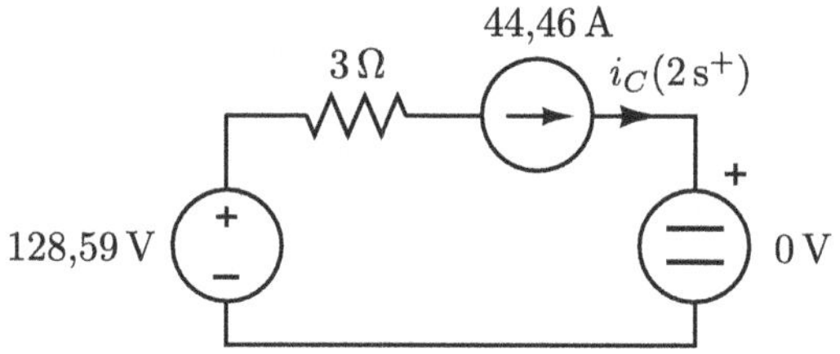

Fig. 2.78 Circuit at $t{=}2^+$

From the circuit in Fig. 2.78, the current $i_C(2\,s^+)$ can be obtained and, consequently, the value of $u'_C(2\,s^+)$ is calculated as:

$$u'_C(2\,s^+) = \frac{i_C(2\,s^+)}{C} = \frac{44{,}46}{3} \approx 14{,}82\,\text{V/s}$$

The steady-state voltage $u_C^p(t)$ is subsequently calculated. Since only an ac-type source of excitation is present, the resulting circuit will be resolved in ac steady-state regime. First, the circuit is obtained in the phasor domain, substituting the resistor, inductor and capacitor for their impedance:

$$\overline{Z}_R = 3\,\Omega \;\; ; \;\; \overline{Z}_L = j\,\Omega \;\; ; \;\; \overline{Z}_C = \frac{-j}{\omega C} = \frac{-j}{3}\,\Omega$$

and the voltage source is replaced by its corresponding phasor:

$$\overline{U}_g = 100\angle 0^\circ\,\text{V}$$

This way the circuit in the phasor domain is that shown in Fig. 2.79.

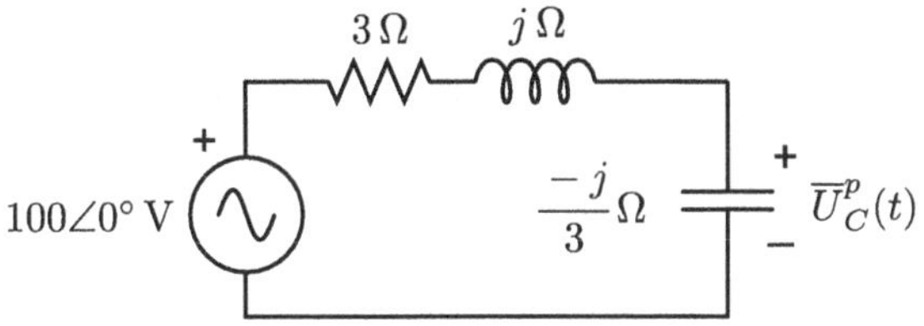

Fig. 2.79 Circuit in the phasor domain

From the circuit in Fig. 2.79, $\overline{U}^p_C$ can be obtained as:

$$\overline{U}^p_C = 100\angle 0^\circ \cdot \frac{-j/3}{3 + j - (j/3)} \approx 10{,}85\angle - 102{,}53^\circ \text{ V}$$

Once the variable of interest has been calculated in the phasor domain, the corresponding expression in the time domain is obtained:

$$u^p_C(t) = \sqrt{2} \cdot 10{,}85 \sin(t - 102{,}53^\circ) \text{ V}$$

The natural response in a second order circuit is characterized by the value of the damping coefficient, α, and that of the resonant frequency, ω_0. Being a series RLC circuit, the values of these parameters are obtained based on R, L and C according to:

$$\alpha = \frac{R}{2L} = \frac{3}{2 \cdot 1} = 1{,}5 \text{ s}^{-1} \ ; \ \omega_0 = \frac{1}{\sqrt{LC}} = \frac{1}{\sqrt{1 \cdot 3}} \approx 0{,}577 \text{ rad s}^{-1}$$

Since $\alpha > \omega_0$, the natural response is overdamped, and considering that the transient starts at $t=2$ s, then:

$$u^n_C(t) = K_1 \cdot e^{s_1(t-2)} + K_2 \cdot e^{s_2(t-2)}$$

where

$$s_1 = -\alpha + \sqrt{\alpha^2 - \omega_0^2} = -1{,}5 + \sqrt{1{,}5^2 - (0{,}577)^2} \approx -0{,}115 \text{ s}^{-1}$$
$$s_2 = -\alpha - \sqrt{\alpha^2 - \omega_0^2} = -1{,}5 - \sqrt{1{,}5^2 - (0{,}577)^2} \approx -2{,}885 \text{ s}^{-1}$$

The complete response of the capacitor voltage is obtained as the sum of the steady-state response and the natural response, yielding:

$$u_C(t) = u^p_C(t) + u^n_C(t) = \sqrt{2} \cdot 10{,}85 \sin(t - 102{,}53^\circ) + K_1 \cdot e^{-0{,}115(t-2)} + K_2 \cdot e^{-2{,}885(t-2)}$$

The values of the constants K_1 and K_2 are calculated from the initial conditions:

$$u_C(2\,\mathrm{s}^+) = 0 = \sqrt{2} \cdot 10{,}85 \sin\left(\frac{2 \cdot 360^\circ}{2\pi} - 102{,}53^\circ\right) + K_1 + K_2$$

$$u'_C(2\,\mathrm{s}^+) = 14{,}82 = \sqrt{2} \cdot 10{,}85 \cos\left(\frac{2 \cdot 360^\circ}{2\pi} - 102{,}53^\circ\right) - 0{,}115K_1 - 2{,}885K_2$$

Resulting in the following system of equations:

$$-3{,}21 = K_1 + K_2$$
$$-0{,}185 = -0{,}115K_1 - 2{,}885K_2$$

The resolution of the system of equations is:

$$K_1 \approx -3{,}41\,\mathrm{V} \;\; ; \quad K_2 \approx 0{,}2\,\mathrm{V}$$

Finally, for t>2 s, the complete response is given by:

$$u_C(t) = \sqrt{2} \cdot 10{,}85 \sin(t - 102{,}53^\circ) - 3{,}41 \cdot \mathrm{e}^{-0{,}115(t-2)} + 0{,}2 \cdot \mathrm{e}^{-2{,}885(t-2)}\,\mathrm{V}$$

The time evolution of $u_C(t)$ has been graphically represented in Fig. 2.80.

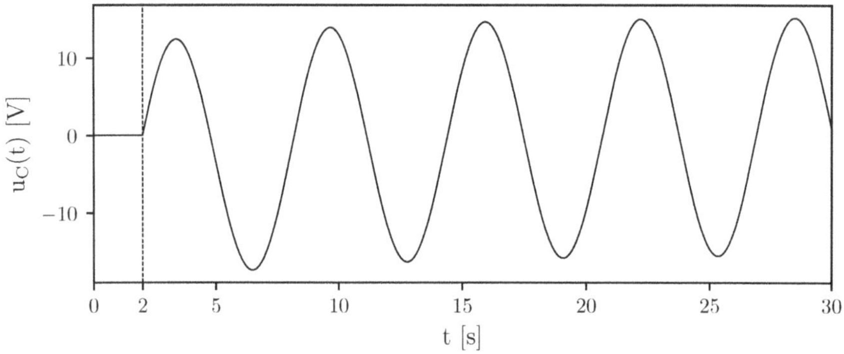

Fig. 2.80 Time evolution of $u_C(t)$

2.4.13 Underdamped Parallel RLC Under DC Supply

In the circuit represented in Fig. 2.81 the switch closes at t=0. Knowing that $u_C(0^-)$=0 V and $i_L(0^-)$=5 A, calculate the current $i_R(t)$ for t>0.

Solution. First, the initial conditions of the variable under study and its derivative, $u_R(0^+)$ and $u'_R(0^+)$, will be calculated. For this purpose, the circuit at t=0$^+$ and the

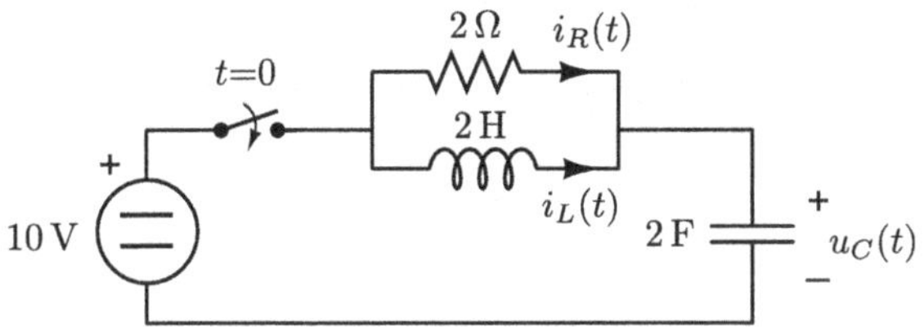

Fig. 2.81 Circuit considered in the problem under study

derivative circuit at t=0$^+$ are analyzed. The first one is obtained by substituting the capacitor for a voltage source of value $u_C(0^+)$ and the inductor for a current source of value $i_L(0^+)$. Except for an impulse-type response, the capacitor voltage and the inductor current do not change from t=0$^-$ to t=0$^+$. Therefore:

$$i_L(0^+) = i_L(0^-) = 5\,\text{A} \ ; \ \ u_C(0^+) = u_C(0^-) = 0\,\text{V}$$

This way, the circuit at t=0$^+$ is that shown in Fig. 2.82.

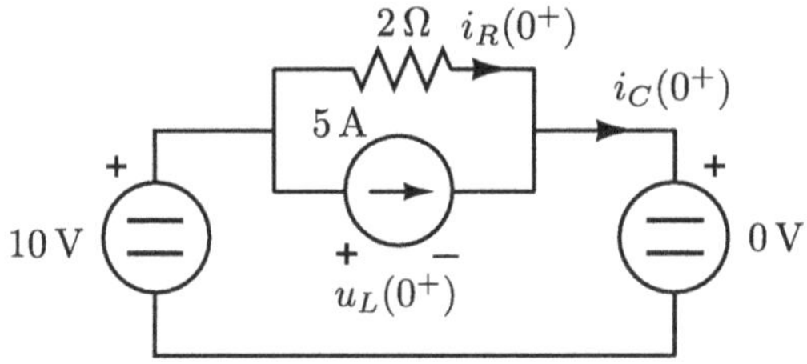

Fig. 2.82 Circuit at t=0$^+$

From the circuit in Fig. 2.82, $i_R(0^+)$ can be obtained as:

$$i_R(0^+) = \frac{10}{2} = 5\,\text{A}$$

On the other hand, the voltage $u_L(0^+)$ and the current $i_C(0^+)$ can be obtained and, consequently, the values of $i'_L(0^+)$ and $u'_C(0^+)$:

$$i'_L(0^+) = \frac{u_L(0^+)}{L} = \frac{10-0}{2} = 5\,\text{A/s}$$

$$u'_C(0^+) = \frac{i_C(0^+)}{C} = \frac{5+5}{2} = 5\,\text{V/s}$$

The derivative circuit at t=0$^+$ is subsequently analyzed, which is obtained by substituting the capacitor for a voltage source of value $u'_C(0^+)$, the inductor for a current source of value $i'_L(0^+)$ and the source of excitation by a source whose value is the derivative of the function at $t=0^+$. For the case of the dc source in this circuit, its derivative is zero and it is replaced by a short circuit. Thus, the derivative circuit at $t=0^+$ is that shown in Fig. 2.83.

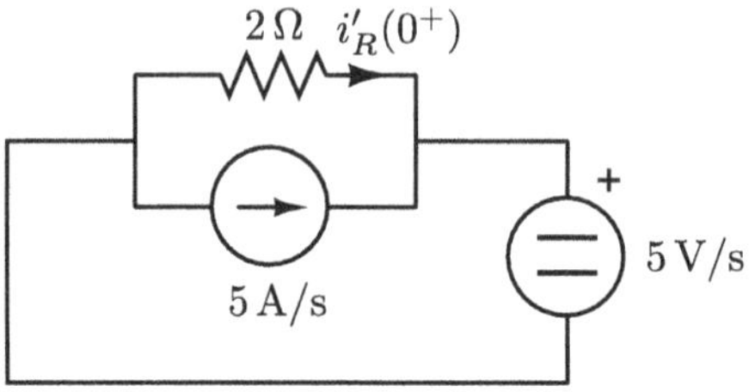

Fig. 2.83 Derivative circuit at $t=0^+$

From the circuit in Fig. 2.83, the value of $i'_R(0^+)$ is obtained:

$$i'_R(0^+) = \frac{-5}{2} = -2{,}5 \text{ A/s}$$

The steady-state current $i^p_R(t)$ is calculated in the sequel. Since only a dc-type source of excitation is present, the circuit in dc steady-state regime shown in Fig. 2.84 will be solved, where the inductor has been replaced by a short circuit and the capacitor by an open circuit.

From the circuit in Fig. 2.84, the value of $i^p_R(t)$ can be obtained:

$$i^p_R(t) = 0 \text{ A}$$

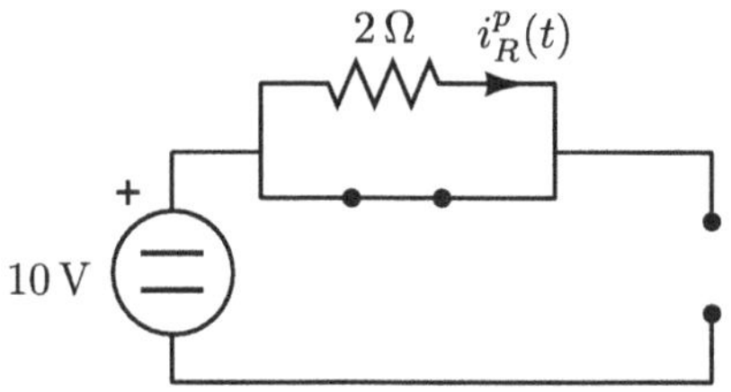

Fig. 2.84 Circuit in dc steady-state regime

The natural response in a second order circuit is characterized by the value of the damping coefficient, α, and that of the resonant frequency, ω_0. Being a parallel RLC circuit, the values of these parameters are obtained based on R, L and C according to:

$$\alpha = \frac{G}{2C} = \frac{1/2}{2 \cdot 2} = 0,125 \text{ s}^{-1} \quad ; \quad \omega_0 = \frac{1}{\sqrt{LC}} = \frac{1}{\sqrt{2 \cdot 2}} = 0{,}5 \text{ rad s}^{-1}$$

where $G=1/R$.

To check that it is a parallel RLC circuit, the passive circuit is analyzed, with the voltage source canceled or short-circuited, given that the natural response is independent of the excitation sources in the circuit.

Since $\alpha < \omega_0$, the natural response is underdamped:

$$i_R^n(t) = K\mathrm{e}^{-\alpha t} \sin(\omega_d t + \varphi)$$

where:

$$\omega_d = \sqrt{\omega_0^2 - \alpha^2} = \sqrt{0{,}5^2 - 0{,}125^2} \approx 0{,}484\,\mathrm{rad\,s^{-1}}$$

The complete response of the circuit is obtained as the sum of the steady-state response and the natural response, yielding:

$$i_R(t) = i_R^p(t) + i_R^n(t) = 0 + K\mathrm{e}^{-0{,}125t} \sin(0{,}484t + \varphi)$$

The values of the constants K and φ are calculated from the initial conditions:

$$i_R(0^+) = 5 = K \sin\varphi$$
$$i'_R(0^+) = -2{,}5 = -0{,}125K \sin\varphi + 0{,}484\,K \cos\varphi$$

To solve the system of equations, the first equation is substituted in the second one, resulting in:

$$-2{,}5 = -0{,}125 \cdot 5 + 0{,}484\,K \cos\varphi \quad \rightarrow \quad K \cos\varphi \approx -3{,}87$$

The value of φ is calculated below:

$$\frac{K \sin\varphi}{K \cos\varphi} = \tan\varphi = \frac{5}{-3{,}87} \quad \rightarrow \quad \varphi \approx -52{,}26^\circ$$

Once the value of φ is obtained, K is calculated as:

$$K = \frac{5}{\sin\varphi} = \frac{5}{\sin(-52{,}26^\circ)} \approx -6{,}32\,\mathrm{A}$$

So:

$$i_R(t) = -6{,}32 \cdot \mathrm{e}^{-0{,}125t} \sin(0{,}484t - 52{,}26^\circ)\ \mathrm{A}$$

The time evolution of $i_R(t)$ has been graphically represented in Fig. 2.85.

2.4.14 Underdamped Parallel RLC Under DC Supply and Impulse-Type Response

The circuit represented in Fig. 2.86 is in steady state when, at $t{=}0$, the switch is closed. Calculate the current $i_L(t)$ for $t{>}0$, given that $C_1{=}3\,\mathrm{F}$ and $C_2{=}1\,\mathrm{F}$.

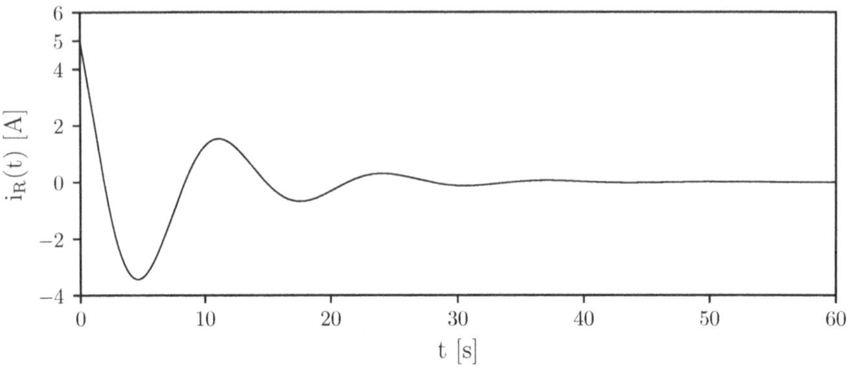

Fig. 2.85 Time evolution of $i_R(t)$

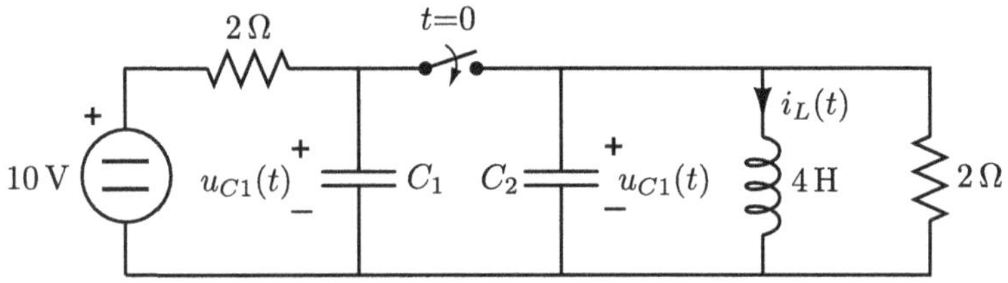

Fig. 2.86 Circuit considered in the problem under study

Solution. First, the voltage will be obtained in each capacitor before the switch closes, that is, at t=0$^-$. For this purpose, it will be considered that the circuit is in steady-state regime. In this situation it can be verified that:

$$u_{C1}(0^-) = 10\,\text{V} \ ; \ u_{C2}(0^-) = 0\,\text{V}$$

Therefore, before the switch closes, the capacitors are charged with different voltages. When the switch is closed, both capacitors are in parallel. Immediately after the closing, the capacitors are subjected to the same voltage at the expense of producing a current pulse. In this instantaneous process, the total charge stored in both capacitors does not change, which allows obtaining the common voltage according to:

$$u_{C1}(0^+) = u_{C2}(0^+) = \frac{C_1 \cdot u_{C1}(0^-) + C_2 \cdot u_{C2}(0^-)}{C_1 + C_2} = \frac{3 \cdot 10 + 1 \cdot 0}{3+1} = \frac{30}{4}\,\text{V}$$

At t=0$^+$, the capacitors are connected in parallel with an initial voltage of 30/4 V. In this situation, both capacitors can be replaced by a single capacitor whose equivalent capacity is

$$C_{eq} = C_1 + C_2 = 3 + 1 = 4\,\text{F}$$

and initial voltage of 30/4 V. This way, the circuit is that shown in Fig. 2.87, where

$$u_{Ceq}(0^+) = \frac{30}{4}\,\text{V}$$

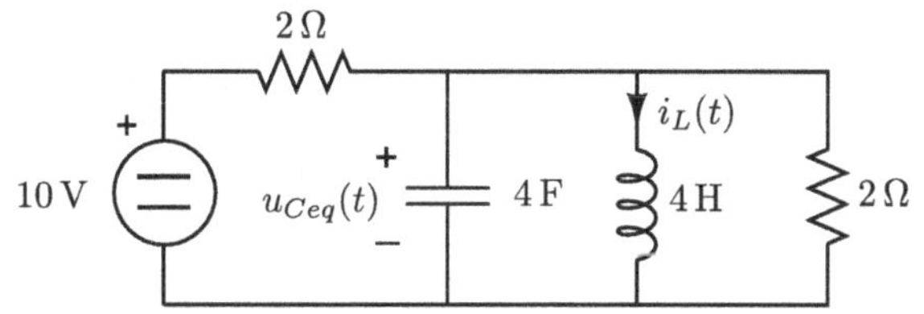

Fig. 2.87 Circuit for $t>0^+$

Then, the initial conditions of the variable under study and its derivative, $i_L(0^+)$ and $i'_L(0^+)$, will be calculated. For this purpose, the circuit at $t=0^+$ is analyzed, which is obtained by substituting the capacitor for a voltage source of value $u_{Ceq}(0^+)$ and the inductor for a current source of value $i_L(0^+)$. Before the switch closes, no current flows through the inductor, as the circuit is in steady state and there is no excitation. Except for an impulse-type response, the inductor current does not change from $t=0^-$ to $t=0^+$. Therefore:

$$i_L(0^+) = i_L(0^-) = 0\,\text{A}$$

This way, the circuit at $t=0^+$ is that shown in Fig. 2.88.

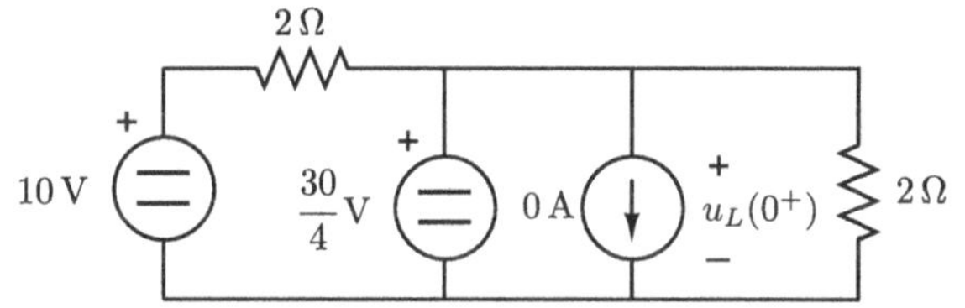

Fig. 2.88 Circuit at $t=0^+$

From the circuit in Fig. 2.88, the voltage $u_L(0^+)$ can be obtained and, consequently, the value of $i'_L(0^+)$ is calculated as:

$$i'_L(0^+) = \frac{u_L(0^+)}{L} = \frac{30/4}{4} = \frac{30}{16}\,\text{A/s}$$

The steady-state current $i_L^p(t)$ is subsequently calculated. Since only a dc-type source of excitation is present, the circuit in steady-state dc regime shown in Fig. 2.89 will be solved, where the inductor has been replaced by a short circuit and the capacitor by an open circuit.

From the circuit in Fig. 2.89, the value of $i_L^p(t)$ is obtained:

$$i_L^p(t) = \frac{10}{2} = 5\,\mathrm{A}$$

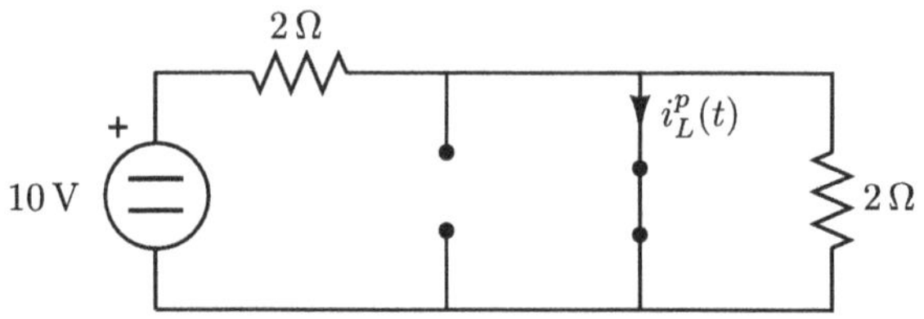

Fig. 2.89 Circuit in steady-state dc regime

The natural response in a second order circuit is characterized by the value of the damping coefficient, α, and that of the resonant frequency, ω_0. To verify that it is a parallel RLC circuit, the passive circuit will be analyzed, which is obtained by canceling the excitation sources, as shown in Fig. 2.90.

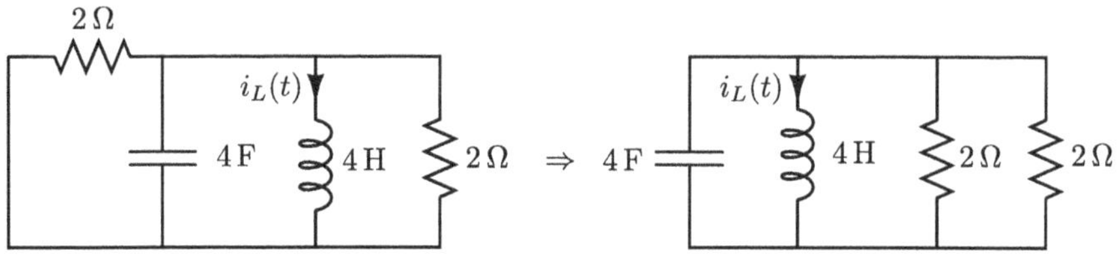

Fig. 2.90 Passive circuit

From the circuit in Fig. 2.90 the equivalent resistance can be calculated:

$$R_{eq} = \frac{2 \cdot 2}{2+2} = 1\,\Omega$$

Being a parallel RLC circuit, the values of α and ω_0 are:

$$\alpha = \frac{G_{eq}}{2C} = \frac{1}{2 \cdot 4} = 0{,}125\,\mathrm{s}^{-1} \;\; ; \;\; \omega_0 = \frac{1}{\sqrt{LC}} = \frac{1}{\sqrt{4 \cdot 4}} = 0{,}25\,\mathrm{rad\,s}^{-1}$$

where $G_{eq}=1/R_{eq}$.

Since $\alpha < \omega_0$, the natural response is underdamped:

$$i_L^n(t) = K\mathrm{e}^{-\alpha t}\sin(\omega_d t + \varphi)$$

where:

$$\omega_d = \sqrt{\omega_0^2 - \alpha^2} = \sqrt{0{,}25^2 - 0{,}125^2} \approx 0{,}2165\,\mathrm{rad\,s}^{-1}$$

The complete response of the circuit is obtained as the sum of the steady-state response plus the natural response, yielding:

$$i_L(t) = i_L^p(t) + i_L^n(t) = 5 + K\mathrm{e}^{-0,125t} \sin(0{,}2165t + \varphi)$$

The values of the constants K and φ are calculated from the initial conditions:

$$i_L(0^+) = 0 = 5 + K \sin \varphi$$
$$i_L'(0^+) = \frac{30}{16} = -0{,}125K \sin\varphi + 0{,}2165\, K \cos\varphi$$

To solve the system of equations, the first equation is substituted in the second one, resulting in:

$$\frac{30}{16} = -0{,}125 \cdot (-5) + 0{,}2165\, K \cos\varphi \quad \rightarrow \quad K \cos\varphi \approx 5{,}77$$

The value of φ is calculated below:

$$\frac{K \sin\varphi}{K \cos\varphi} = \tan\varphi = \frac{-5}{5{,}77} \quad \rightarrow \quad \varphi \approx -40{,}91^\circ$$

Once the value of φ is obtained, K is calculated as:

$$K = \frac{-5}{\sin\varphi} = \frac{-5}{\sin(-40{,}91^\circ)} \approx 7{,}64\ \mathrm{A}$$

So:

$$i_L(t) = 5 + 7{,}64 \cdot \mathrm{e}^{-0,125t} \sin(0{,}2165t - 40{,}91^\circ)\ \mathrm{A}$$

The time evolution of $i_L(t)$ has been graphically represented in Fig. 2.91.

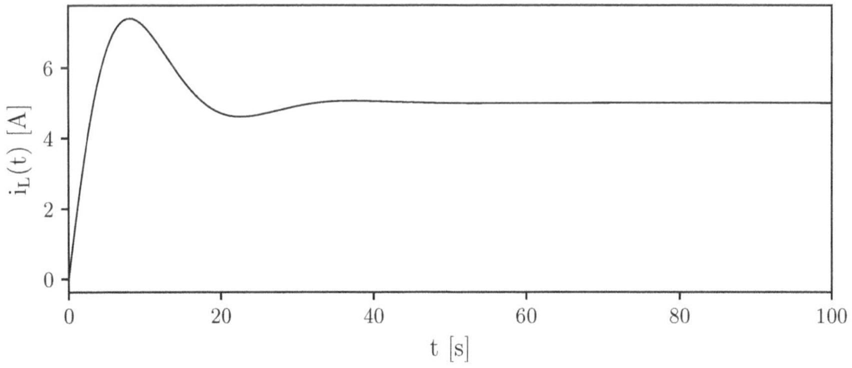

Fig. 2.91 Time evolution of $i_L(t)$

2.4.15 Obtaining the Differential Equation

In the circuit represented in Fig. 2.92, deduce the differential equation of the current $i(t)$ as a function of $e_g(t)$, R_1, R_2, L and C.

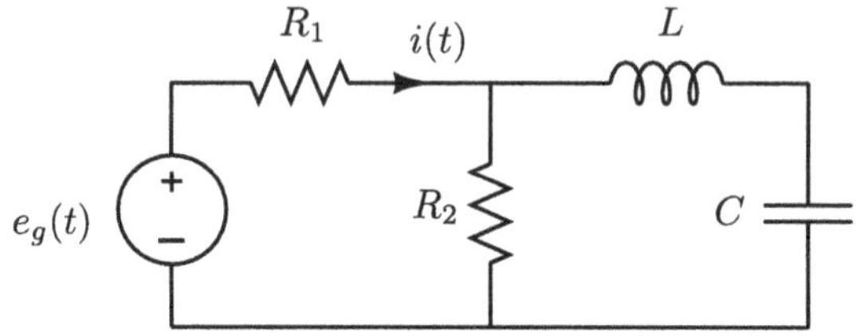

Fig. 2.92 Circuit considered in the problem under study

Solution. To simplify the notation, additional variables have been included, as shown in Fig. 2.93.

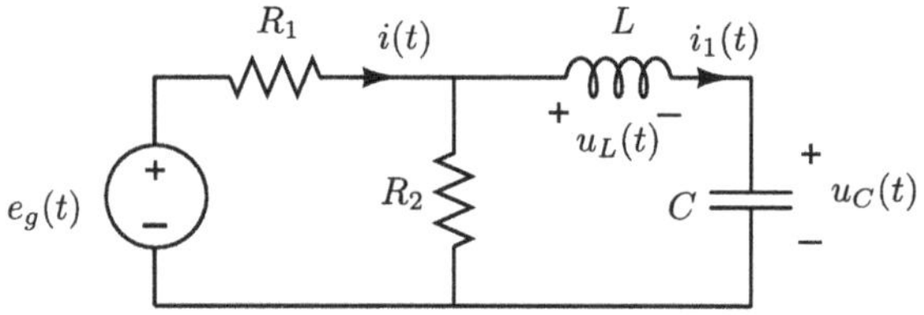

Fig. 2.93 Circuit with additional variables

Kirchhoff's voltage law is applied to the loop formed by $e_g(t)$, R_1, L and C, resulting in:

$$e_g(t) = R_1 \cdot i(t) + u_L(t) + u_C(t) \tag{2.66}$$

Given that

$$u_L = L \cdot \frac{di_1(t)}{dt}$$

then (2.66) is modified, yielding:

$$e_g(t) = R_1 \cdot i(t) + L \cdot \frac{di_1(t)}{dt} + u_C(t) \tag{2.67}$$

The differentiation of Eq. (2.67) results in:

$$\frac{de_g(t)}{dt} = R_1 \cdot \frac{di(t)}{dt} + L \cdot \frac{d^2i_1(t)}{dt^2} + \frac{du_C(t)}{dt}$$

and considering that

$$i_1 = C \cdot \frac{du_C(t)}{dt}$$

then

$$\frac{de_g(t)}{dt} = R_1 \cdot \frac{di(t)}{dt} + L \cdot \frac{d^2 i_1(t)}{dt^2} + \frac{i_1(t)}{C}$$

Finally, a relation between $i(t)$ and $i_1(t)$ is obtained by applying Kirchhoff's voltage law to the loop formed by the elements $e_g(t)$, R_1 and R_2

$$e_g(t) = R_1 \cdot i(t) + R_2 \cdot (i(t) - i_1(t)) \Rightarrow i_1(t) = \frac{(R_1 + R_2) \cdot i(t) - e_g(t)}{R_2}$$

Substituting and rearranging terms results in:

$$\frac{d^2 i(t)}{dt^2} + \frac{R_1 R_2}{L(R_1 + R_2)} \cdot \frac{di(t)}{dt} + \frac{i(t)}{LC} = \frac{1}{(R_1 + R_2)} \cdot \left(\frac{d^2 e_g(t)}{dt^2} + \frac{R_2}{L} \cdot \frac{de_g(t)}{dt} + \frac{e_g(t)}{LC(R_1 + R_2)} \right)$$

2.4.16 Obtaining the Differential Equation

The circuit represented in Fig. 2.94 is in steady state before the switch changes position. It is requested:

(a) Obtain the differential equation to calculate $u(t)$ for $t>0$.
(b) From the differential equation obtained in the previous section, calculate the damping coefficient and the resonance pulsation, and based on these values, deduce what type of natural response of the circuit.
(c) Calculate $u(0^+)$ and $u'(0^+)$.

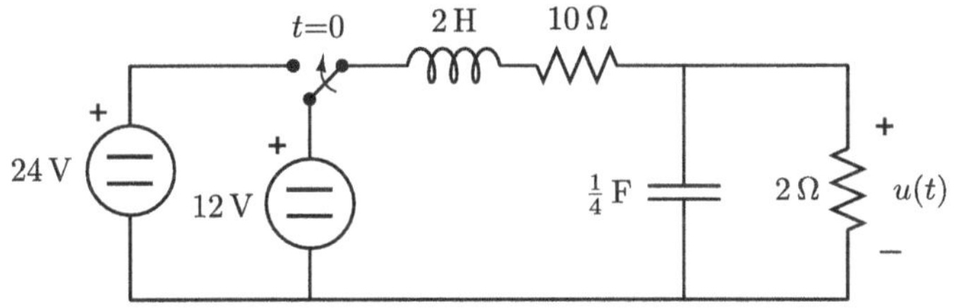

Fig. 2.94 Circuit considered in the problem under study

Solution. (a) For $t>0$, the resulting circuit is that shown in Fig. 2.95, where additional variables have been included to simplify the notation.
Applying Kirchhoff's voltage law to the loop formed by the inductor, resistor and capacitor results in:

$$24 = u_L(t) + 10 \cdot i(t) + u(t) \tag{2.68}$$

Given that

$$u_L(t) = 2 \cdot \frac{di(t)}{dt}$$

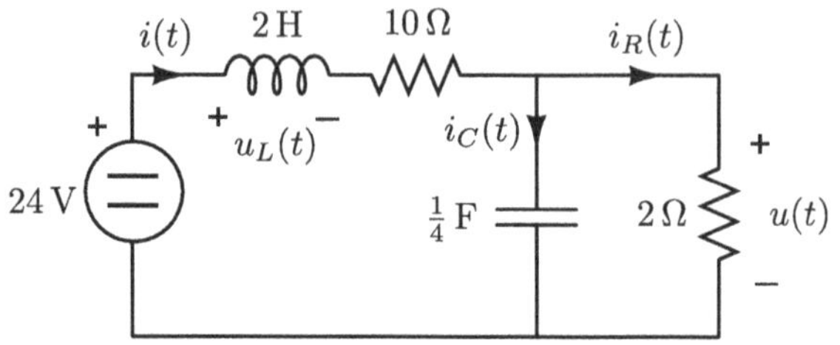

Fig. 2.95 Circuit for $t > 0$ s

the Eq. (2.68) is modified

$$24 = 2 \cdot \frac{di(t)}{dt} + 10 \cdot i(t) + u(t) \tag{2.69}$$

Applying Kirchhoff's current law:

$$i(t) = i_C(t) + i_R(t) \tag{2.70}$$

Given that

$$u(t) = 2 \cdot i_R(t) \ ; \ \ i_C(t) = \frac{1}{4} \cdot \frac{du(t)}{dt}$$

the Eq. (2.70) is modified

$$i(t) = \frac{1}{4} \cdot \frac{du(t)}{dt} + \frac{u(t)}{2} \tag{2.71}$$

Substituting (2.71) in (2.69) results in:

$$24 = 2 \cdot \frac{d}{dt}\left(\frac{1}{4} \cdot \frac{du(t)}{dt} + \frac{u(t)}{2}\right) + 10 \cdot \left(\frac{1}{4} \cdot \frac{du(t)}{dt} + \frac{u(t)}{2}\right) + u(t)$$

Operating and rearranging terms:

$$\frac{d^2u(t)}{dt^2} + 7 \cdot \frac{du(t)}{dt} + 12 \cdot u(t) = 48 \tag{2.72}$$

(b) Analyzing the differential equation (2.72), the damping coefficient, α and the resonant frequency, ω_0 are obtained:

$$2\alpha = 7 \ \Rightarrow \ \alpha = \frac{7}{2} = 3{,}5\,\text{s}^{-1}$$

$$\omega_0^2 = 12 \ \Rightarrow \ \omega_0 = \sqrt{12} \approx 3{,}46\,\text{rad}\,\text{s}^{-1}$$

Since $\alpha > \omega_0$, the natural response is overdamped.

(c) To calculate $u(0^+)$ and $u'(0^+)$, the circuit at t=0^- will be analyzed. According to the problem statement, before the switch changes its position, the circuit is in steady state. Since only a dc-type source of excitation is present, the corresponding circuit in dc steady-state regime will be solved, which is obtained by replacing the inductor with a short circuit and the capacitor with an open circuit, as shown in Fig. 2.96. According to this circuit:

$$i_L(0^-) = \frac{12}{10 \mid 2} = 1\,\text{A} \;;\;\; u_C(0^-) = 2 \cdot 1 = 2\,\text{V}$$

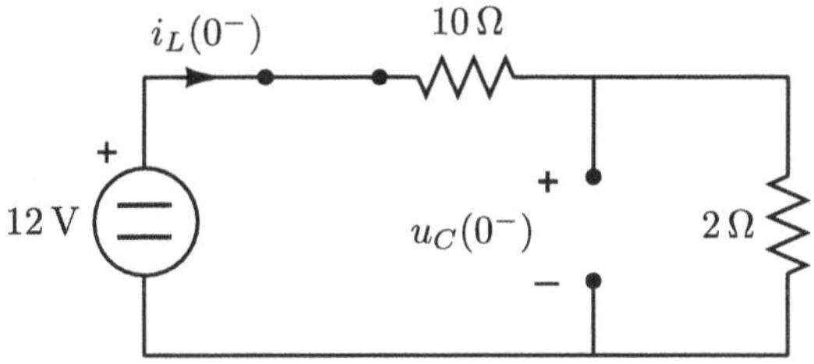

Fig. 2.96 Circuit at t=0^- s

Except for an impulse-type response, the capacitor voltage and the inductor current do not change from t=0^- to t=0^+. Therefore:

$$i_L(0^+) = i_L(0^-) = 1\,\text{A} \;;\;\; u_C(0^+) = u_C(0^-) = 2\,\text{V}$$

To obtain $u(0^+)$, the circuit at t=0^+ is analyzed, which is obtained by substituting the capacitor for a voltage source of value $u_C(0^+)$ and the inductor by a current source of value $i_L(0^+)$. This way, the circuit at t=0^+ is that shown in Fig. 2.97.

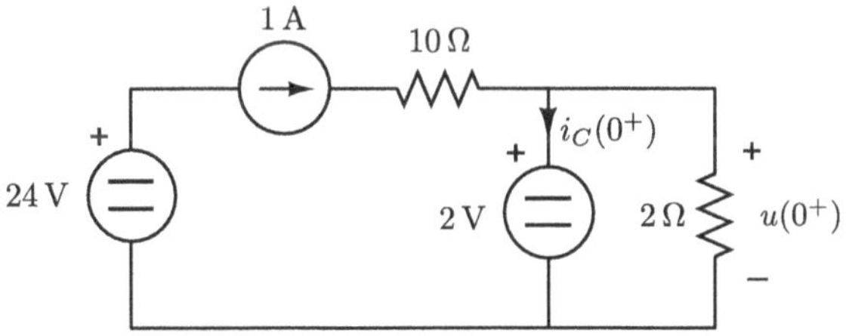

Fig. 2.97 Circuit at t=0^+

From the circuit in Fig. 2.97, the values of $u(0^+)$ and $i_C(0^+)$ can be obtained

$$u(0^+) = 2\,\text{V}$$

$$i_C(0^+) = 1 - \frac{u(0^+)}{2} = 1 - \frac{2}{2} = 0\,\text{A}$$

With the value of $i_C(0^+)$ and considering that $u(t)=u_C(t)$, then:

$$u'(0^+) = u'_C(0^+) = \frac{i_C(0^+)}{C} = \frac{2}{1/4} = 8\ \text{V/s}$$

Chapter 3
Laplace Transform Analysis

Abstract This chapter shows the application of the Laplace transform to the resolution of electrical circuits, beginning with a theoretical introduction of the concepts required to correctly address each of the subsequent problems. A total of 19 fully solved problems with explanatory comments are included.

3.1 Introduction

Although the Laplace transform is generally applied to the analysis of any electrical circuit, it particularly facilitates the study of transient regimes in linear networks of any order. Additionally, this transform can also be applied to any type of excitation, as long as the network is linear and preferably with time-invariant parameters. Essentially, the Laplace transform is capable of transforming differential equations into algebraic equations, in a similar way to that of the phasor techniques studied for the resolution of ac circuits.

3.2 Definition

Let $f(t)$ be a time function. The **Laplace transform** of $f(t)$ is a s-domain function defined as:

$$\mathscr{L}[f(t)] = F(s) \triangleq \int_{0^-}^{\infty} f(t)\mathrm{e}^{st}dt \tag{3.1}$$

where the variable s is the complex frequency and, equal to

$$s = \sigma + j\omega$$

and whose unit is s^{-1}.

A. Bachiller Soler et al., *Solved Problems for Transient Electrical Circuits*,
Lecture Notes in Electrical Engineering 809,
https://doi.org/10.1007/978-3-030-88144-3_3

Regarding the Laplace transform, the following observations can be made:

- It is not time-dependent.
- It does not consider the value of $f(t)$ for $t < 0^-$.
- It can be considered as a change of variables, from time to complex frequency.

3.3 Main Properties and Theorems

The main properties and theorems related to the Laplace transform are listed below, many of which will be used in the analysis of electrical circuits.

1. Linearity:

$$\mathscr{L}[k_1 f_1(t) + k_2 f_2(t)] = k_1 \mathscr{L}[f_1(t)] + k_2 \mathscr{L}[f_2(t)] = k_1 F_1(s) + k_2 F_2(s)$$

2. Differentiation:

$$\mathscr{L}\left[\frac{df(t)}{dt}\right] = s\mathscr{L}[f(t)] - f(0^-) = sF(s) - f(0^-)$$

$$\mathscr{L}\left[\frac{d^2 f(t)}{dt^2}\right] = s^2 F(s) - sf(0^-) - f'(0^-)$$

3. Integration:

$$\mathscr{L}\left[\int_{0^-}^{t} f(t)dt\right] = \frac{\mathscr{L}[f(t)]}{s} = \frac{F(s)}{s}$$

4. Differentiation in s-domain:

$$\mathscr{L}[t\, f(t)] = \frac{-dF(s)}{ds}$$

5. Frequency shifting:

$$\mathscr{L}\left[e^{at} f(t)\right] = F(s-a)$$

6. Time shifting:

$$\mathscr{L}\left[f(t-a)\right] = e^{-as} F(s),\ a > 0$$

7. Time scaling:

$$\mathscr{L}\left[f(at)\right] = \frac{1}{a} F\left(\frac{s}{a}\right)$$

8. Initial value theorem:

$$\lim_{t \to 0^+} f(t) = \lim_{s \to \infty} sF(s)$$

9. Final value theorem:

$$\lim_{t \to \infty} f(t) = \lim_{s \to 0} sF(s)$$

3.4 Laplace Transform Pairs

Table 3.1 shows the Laplace transform pairs which are commonly used in the study of electrical circuits.

Table 3.1 Laplace transform pairs

$f(t)$	$F(s)$
k	$\frac{k}{s}$
$\delta(t)$	1
$u(t)$	$\frac{1}{s}$
t^r	$\frac{r!}{s^{r+1}}$
e^{-at}	$\frac{1}{s+a}$
$t^r e^{-at}$	$\frac{r!}{(s+a)^{r+1}}$

$f(t)$	$F(s)$
$\sin(\omega t)$	$\frac{\omega}{s^2+\omega^2}$
$\cos(\omega t)$	$\frac{s}{s^2+\omega^2}$
$\sin(\omega t+\varphi)$	$\frac{s\sin\varphi+\omega\cos\varphi}{s^2+\omega^2}$
$\cos(\omega t+\varphi)$	$\frac{s\cos\varphi-\omega\sin\varphi}{s^2+\omega^2}$
$2Re^{\alpha t}\cos(\beta t+\psi)$	$\frac{R\angle\psi}{[s-(\alpha+j\beta)]}+\frac{R\angle-\psi}{[s-(\alpha-j\beta)]}$

3.5 Application to the Analysis of Electrical Circuits

3.5.1 *Introduction*

A diagram is shown in Fig. 3.1 which summarizes the process of solving a transient problem using classical analysis in the time domain and using the Laplace transform.

In the time domain, first (resp. second) order circuits lead to linear time-invariant first (resp. second) order differential equations. These equations are solved by applying the classical techniques, identifying the natural response and the steady-state response. However, this method becomes more complicated when the excitation sources are not the usual dc or ac sources. Likewise, in second-order circuits that are not "canonical", that is, those which are not series RLC or parallel RLC type, there is no a priori defined method to find the parameters defining the natural response. In these particular cases, and generally, the Laplace transform is one of the methods that can be applied to the resolution of circuits, with no limitation in terms of the type of excitation sources and the circuit order.

In general, the application of the Laplace transform allows transforming a differential equation into an algebraic one of relative simplicity, which can be expressed in the desired form. With the latter results and using the inverse transformation, the complete solution of the initial differential equation is obtained. If this applica-

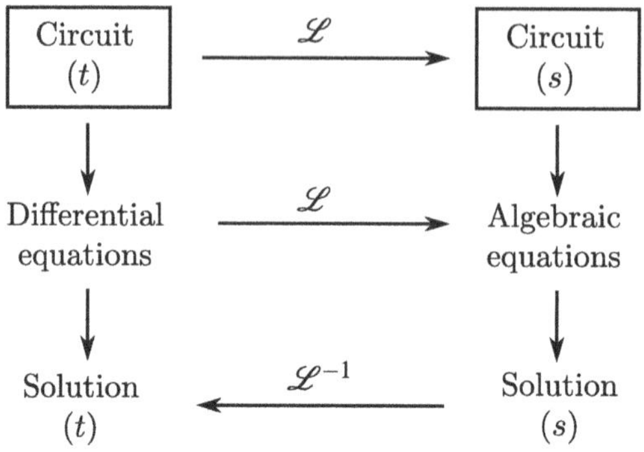

Fig. 3.1 Process of solving a transient problem using classical analysis in the time domain and using the Laplace transform

tion is taken to the extreme, a circuit in the time domain can be transformed to the frequency domain simply by using the Laplace transform with the voltage-current relationships of each of the elements involved in the circuit. Once the circuit is in the Laplace domain, the equations that govern those relationships between voltage and current become algebraic. Obviously, the solution of the circuit, that is, the calculation of one or several variables of interest, will be expressed in the Laplace domain. To obtain this solution in the time domain it will be necessary to use the inverse Laplace transform. However, this process often becomes difficult, being this the bottleneck of this method.

3.5.2 *Voltage-Current Relationship in the s-Domain*

In the sequel, the voltage-current relationship will be obtained in the s-domain for different circuit elements.

3.5.2.1 Resistor

A resistor has been represented in Fig. 3.2 with the corresponding voltage and current references in the time domain.

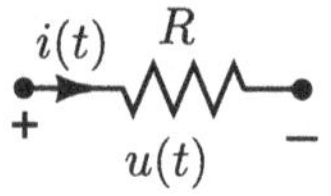

Fig. 3.2 Resistor in the time domain

From the scheme in Fig. 3.2:

$$u(t) = R \cdot i(t) \tag{3.2}$$

Considering that

$$\mathscr{L}[u(t)] = U(s) \tag{3.3}$$

and

$$\mathscr{L}[R \cdot i(t)] = R \cdot \mathscr{L}[i(t)] = R \cdot I(s) \tag{3.4}$$

the following result is obtained:

$$U(s) = R \cdot I(s) \tag{3.5}$$

A resistor has been represented in Fig. 3.3 in the Laplace domain.

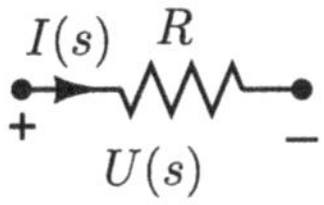

Fig. 3.3 Resistor in the Laplace domain

3.5.2.2 Inductor

An inductor has been represented in Fig. 3.4 with the corresponding voltage and current references in the time domain.

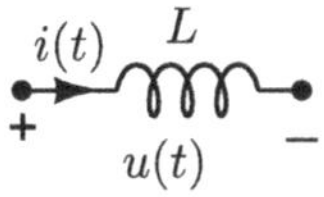

Fig. 3.4 Inductor in the time domain

From the scheme in Fig. 3.4:

$$u(t) = L \cdot \frac{di(t)}{dt} \tag{3.6}$$

Considering that

$$\mathscr{L}[u(t)] = U(s) \tag{3.7}$$

and

$$\mathscr{L}\left[L \cdot \frac{di(t)}{dt}\right] = L \cdot \mathscr{L}\left[\frac{di(t)}{dt}\right] = L \cdot [sI(s) - i(0^-)] = LsI(s) - Li(0^-) \tag{3.8}$$

the following result is obtained:

$$U(s) = LsI(s) - Li(0^-) \tag{3.9}$$

where $i(0^-)$ represents the current through the inductor at $t = 0^-$.

The equivalent circuit of an inductor has been represented in Fig. 3.5 in the Laplace domain.

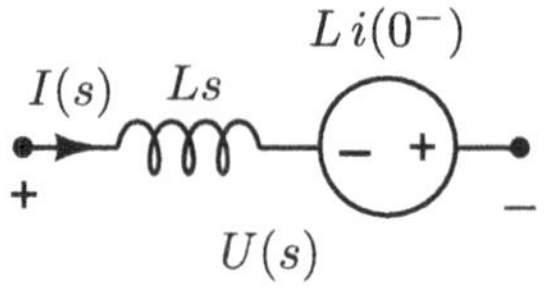

Fig. 3.5 Equivalent circuit of an inductor in the Laplace domain. Model with voltage source

On the other hand, the relationship (3.9) can be expressed as follows:

$$I(s) = \frac{U(s)}{Ls} + \frac{i(0^-)}{s} \tag{3.10}$$

This relationship leads to an alternative equivalent circuit of an inductor in the Laplace domain, as represented in Fig. 3.6.

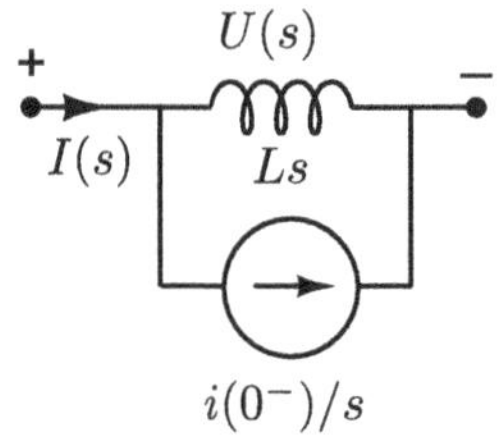

Fig. 3.6 Equivalent circuit of an inductor in the Laplace domain. Model with current source

Depending on the particular circuit to be solved, it will be more adequate to use the model with a voltage source or that with a current source, given that, obviously, both models are equivalent.

3.5.2.3 Capacitor

A capacitor has been represented in Fig. 3.7 with the corresponding voltage and current references in the time domain.

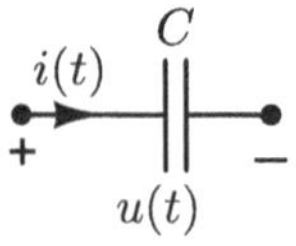

Fig. 3.7 Capacitor in the time domain

From the scheme in Fig. 3.7:

$$i(t) = C \cdot \frac{du(t)}{dt} \tag{3.11}$$

Considering that

$$\mathscr{L}[i(t)] = I(s) \tag{3.12}$$

and

$$\mathscr{L}[C \cdot \frac{du(t)}{dt}] = C \cdot \mathscr{L}[\frac{du(t)}{dt}] = C \cdot [sU(s) - u(0^-)] = CsU(s) - Cu(0^-) \tag{3.13}$$

the following result is obtained:

$$I(s) = CsU(s) - Cu(0^-) \tag{3.14}$$

where $u(0^-)$ represents the capacitor voltage at $t = 0^-$.

In Fig. 3.8 The equivalent circuit of a capacitor has been represented in Fig. 3.8 in the Laplace domain.

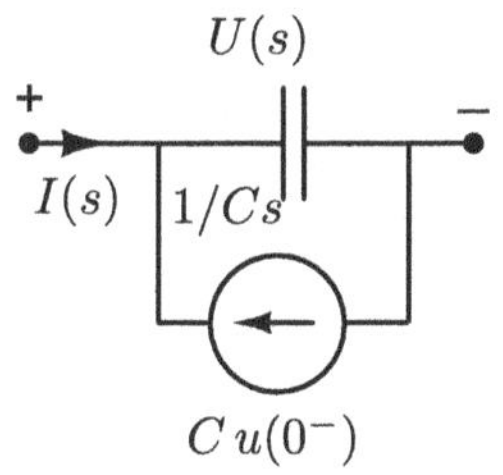

Fig. 3.8 Equivalent circuit of a capacitor in the Laplace domain. Model with current source

On the other hand, the relationship (3.14) can be expressed as follows:

$$U(s) = \frac{I(s)}{Cs} + \frac{u(0^-)}{s} \tag{3.15}$$

This relationship leads to an alternative equivalent circuit of a capacitor in the Laplace domain, as represented in Fig. 3.9.

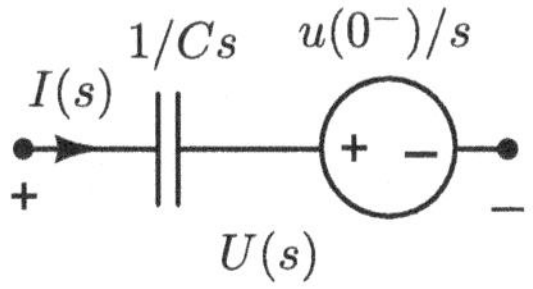

Fig. 3.9 Equivalent circuit of a capacitor in the Laplace domain. Model with voltage source

As mentioned before, depending on the particular circuit under study, it will be more adequate to use the model with a voltage source or that with a current source, given that, obviously, both models are equivalent.

3.5.2.4 Independent Sources

The equivalent circuit of the independent sources in the Laplace domain is obtained with the corresponding Laplace transform of the defining function, as shown in Fig. 3.10.

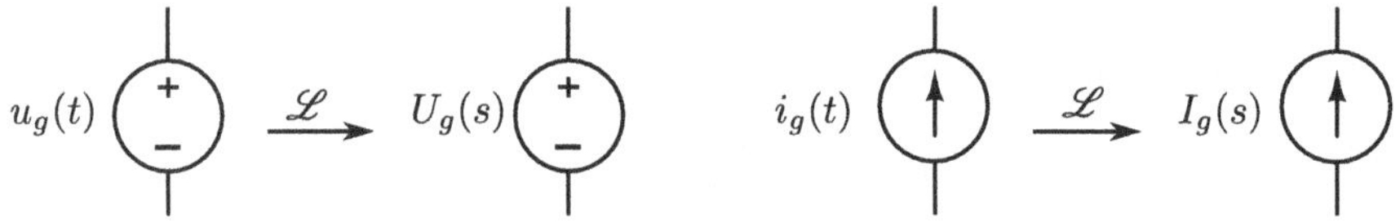

Fig. 3.10 Model of independent sources in the Laplace domain

3.5.2.5 Dependent Sources

In Figs. 3.11 and 3.12, the Laplace transforms of the voltage and current dependent sources are respectively shown.

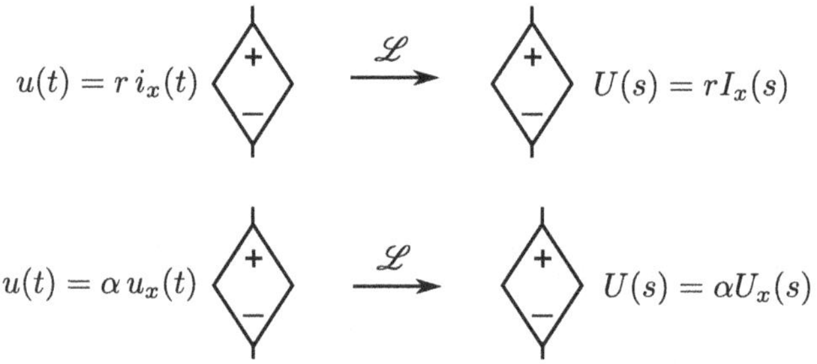

Fig. 3.11 Model of voltage dependent sources in the Laplace domain

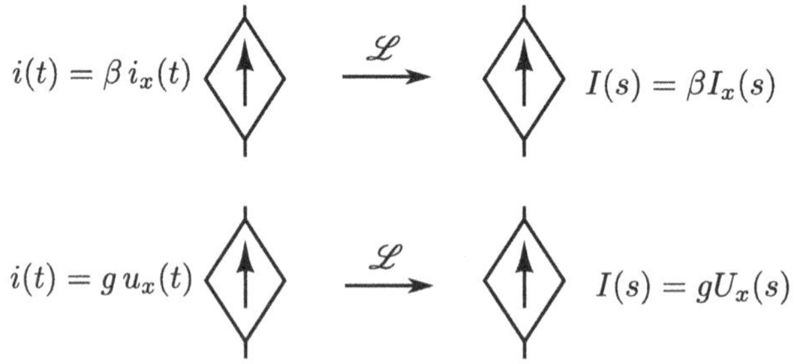

Fig. 3.12 Model of current dependent sources in the Laplace domain

The current $i_x(t)$ and the voltage $u_x(t)$ are the control variables of the dependent sources.

3.5.2.6 Magnetically Coupled Coils

Two magnetically coupled coils are shown in Fig. 3.13 with the corresponding voltage and current references in the time domain.

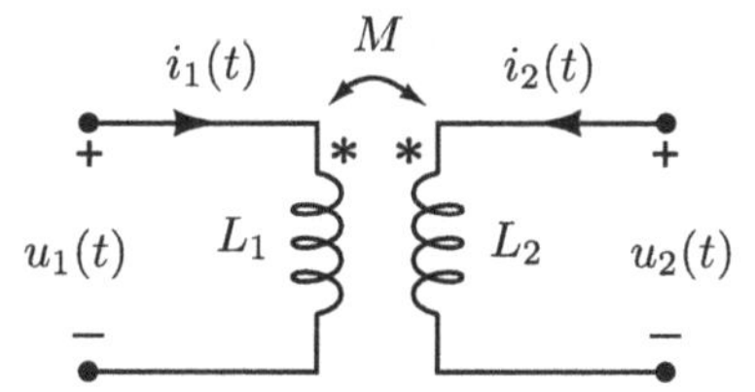

Fig. 3.13 Magnetically coupled coils in the time domain

From the circuit in Fig. 3.13:

$$\begin{aligned} u_1(t) &= L_1 \cdot \frac{di_1(t)}{dt} + M \cdot \frac{di_2(t)}{dt} \\ u_2(t) &= M \cdot \frac{di_1(t)}{dt} + L_2 \cdot \frac{di_2(t)}{dt} \end{aligned} \tag{3.16}$$

Using the Laplace transform in both equations results in:

$$\begin{aligned} U_1(s) &= L_1 s I_1(s) + M s I_2(s) - L_1 i_1(0^-) - M i_2(0^-) \\ U_2(s) &= M s I_1(s) + L_2 s I_2(s) - M i_1(0^-) - L_2 i_2(0^-) \end{aligned} \tag{3.17}$$

where $i_1(0^-)$ and $i_2(0^-)$ represent the currents for each inductor at $t = 0^-$.

The equivalent circuit of two magnetically coupled coils has been represented in Fig. 3.14 in the Laplace domain.

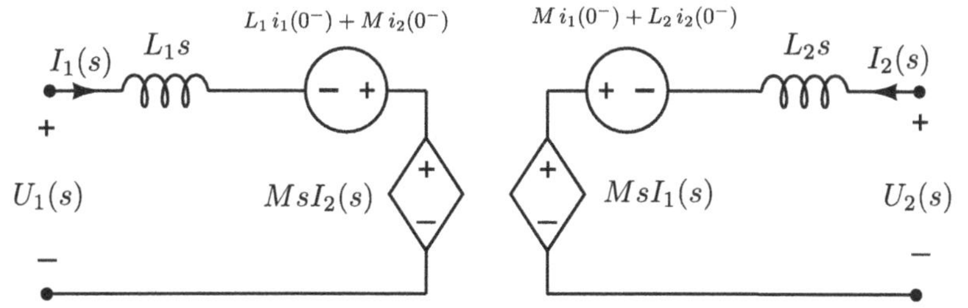

Fig. 3.14 Equivalent circuit of two magnetically coupled coils in the Laplace domain

3.5.3 *Impedance and Admittance*

For any two-terminal passive circuit (without independent sources or initial conditions), the relationship between the applied voltage, $U(s)$, and the flowing current, $I(s)$, is defined as the impedance of the circuit (Fig. 3.15). This definition can be considered a generalization of Ohm's law for resistive circuits in Laplace domain.

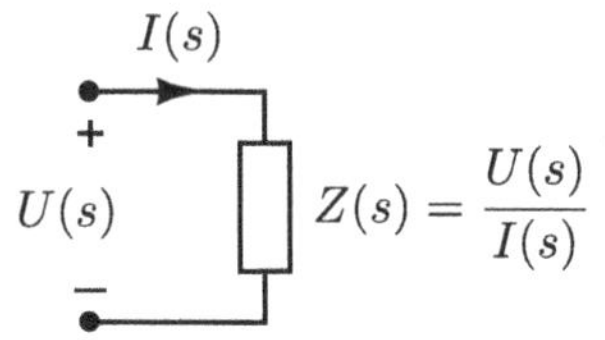

Fig. 3.15 Impedance

Generalizing the case of conductance, the admittance of a two-terminal element is defined as:

$$Y(s) = \frac{1}{Z(s)} \tag{3.18}$$

The SI unit of an impedance is the Ohm [Ω], and that of an admittance is the Siemens [S].

Table 3.2 shows the impedance and the admittance of a resistor, an inductor and a capacitor.

Table 3.2 Impedance and admittance of resistor, inductor and capacitor

Element	**Z(s)**	**Y(s)**
Resistor	R	$1/R$
Inductor	Ls	$1/Ls$
Capacitor	$1/Cs$	Cs

3.5.4 *Kirchhoff's Laws*

Kirchhoff's current law establishes that, in an electric node where k conductors converge, with their currents assumed to be entering the node, it is verified that:

$$\sum_k i_k(t) = 0 \tag{3.19}$$

Applying the Laplace transform results in:

$$\mathscr{L}\left[\sum_k i_k(t)\right] = 0 \Rightarrow \sum_k \mathscr{L}\left[i_k(t)\right] = 0 \Rightarrow \sum_k I_k(s) = 0 \tag{3.20}$$

On the other hand, Kirchhoff's voltage law establishes that, in an electric loop formed by k elements, whose voltages are assumed to have the same direction in which the loop is traversed, it is verified that:

$$\sum_k u_k(t) = 0 \tag{3.21}$$

Applying the Laplace transform results in:

$$\mathscr{L}\left[\sum_k u_k(t)\right] = 0 \Rightarrow \sum_k \mathscr{L}\left[u_k(t)\right] = 0 \Rightarrow \sum_k U_k(s) = 0 \tag{3.22}$$

It can be concluded that Kirchhoff's laws are also satisfied in the Laplace domain.

3.5.5 Methodology of Resolution

The methodology for the resolution of an electrical circuit using the Laplace transform is summarized below.

1. Derivation of the circuit in the frequency domain. For this purpose, each elements of the circuit is represented in the s-domain, using the relationships between voltage and current deduced in the Sect. 3.5.2. Likewise, it will be necessary to consider the inductor currents and the capacitor voltages at instant $t = 0^-$. This process is described in Fig. 3.16, where $u_o(t)$ and $i_o(t)$ represent an output voltage and current in time domain at the determined point of the circuit. Logically, $U_o(s)$ and $I_o(s)$ represent the same outputs in Laplace domain.
2. Formulation and resolution of the equations for the variables under study. For this point, it must be considered that in the s-domain the following techniques and rules can still be applied:
    - The rules for grouping elements.
    - The rules for transforming sources.
    - Kirchhoff's laws.
    - The mesh analysis technique.
    - The node analysis technique.
    - The Thévenin and Norton equivalents.
    - Principle of superposition.
3. Transformation of the obtained results into their equivalents in the time domain (inverse Laplace transform).

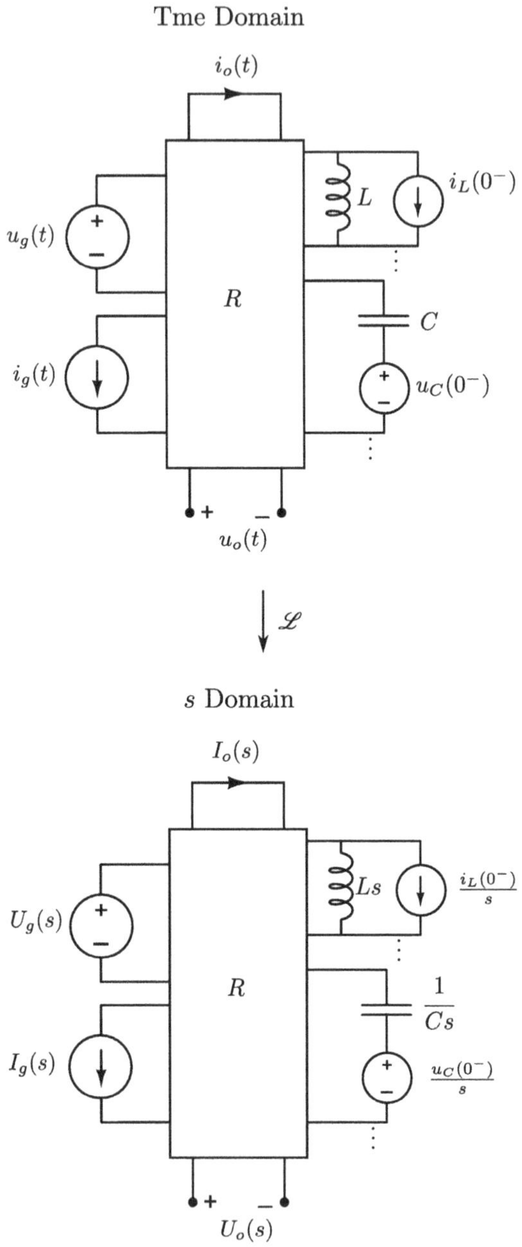

Fig. 3.16 Methodology of resolution using the Laplace transform

Example

Calculate the current $I(s)$, transformed from $i(t)$, which circulates through the circuit represented in Fig. 3.17, given that $u_g(t)$=5 V, R=2 Ω, L=0,1 H, C=0,5 F, $u_C(0^-)$=0 V and $i_L(0^-)$=0 A.

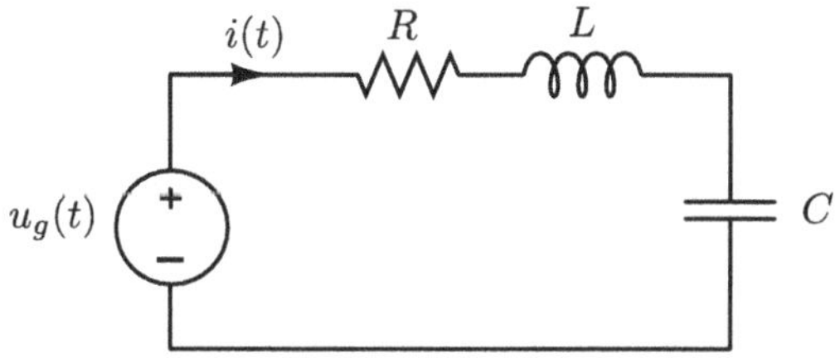

Fig. 3.17 Example circuit

Solution The circuit in the Laplace domain is shown in Fig. 3.18. From the circuit in Fig. 3.18:

$$I(s) = \frac{5/s}{2 + 0,1s + \frac{2}{s}}$$

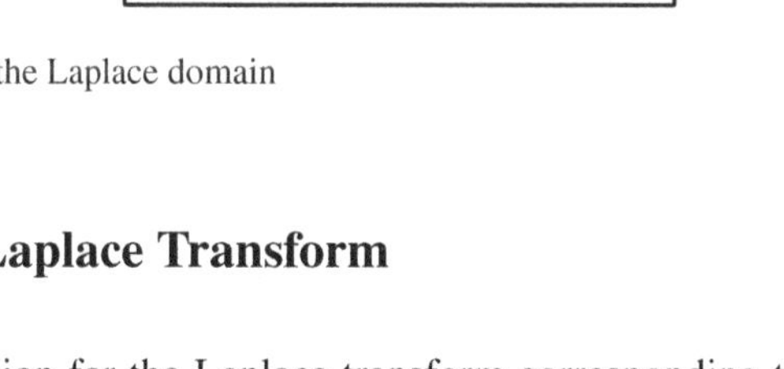

Fig. 3.18 Circuit in the Laplace domain

3.6 Inverse Laplace Transform

A general expression for the Laplace transform corresponding to a variable in the circuit under study is a rational function given by:

$$F(s) = \frac{a_n s^n + a_{n-1}s^{n-1} + \cdots + a_2 s^2 + a_1 s + a_0}{s^m + b_{m-1}s^{m-1} + \cdots + b_2 s^2 + b_1 s + b_0} \tag{3.23}$$

where a_i, b_j are real constants, while n and m are positive integers.

- If $n < m$ then $F(s)$ is a **proper rational function**.
- If $n \geq m$ then $F(s)$ is an **improper rational function**. In this case, $F(s)$ can be expressed as

$$F(s) = Q(s) + P(s) \tag{3.24}$$

  where:

  - $Q(s)$ is a polynomial of degree $n - m$.
  - $P(s)$ it is a proper rational function.

For physically realizable circuits, it is verified that $n \leq m$, so that $F(s)$ can be expressed as

$$F(s) = C + P(s) \tag{3.25}$$

where C is a constant and $P(s)$ is a proper rational function.

Example Calculation of $Q(s)$ and $P(s)$

$$F(s) = \frac{4s^2 + 17s + 16}{s^2 + 4s + 3} = \frac{D(s)}{d(s)} = Q(s) + P(s)$$

Solution

$$F(s) = \frac{D(s)}{d(s)} = c(s) + \frac{r(s)}{d(s)} = 4 + \frac{s+4}{s^2 + 4s + 3}$$

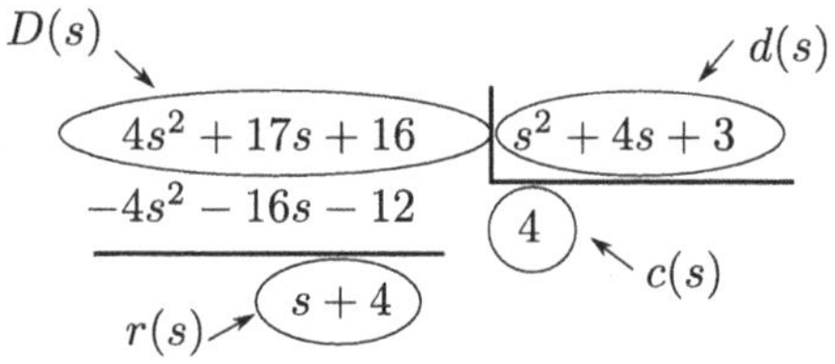

Fig. 3.19 Polynomial division

From the previous results:

$$Q(s) = 4 \;\; ; \quad P(s) = \frac{s+4}{s^2 + 4s + 3}$$

3.6.1 *Inverse Laplace Transform Calculation Methodology*

- The objective is finding $f(t)$ from $F(s)$.
- $Q(s)$ is a polynomial of degree $n - m$, whose inverse Laplace transform is directly obtained.

- $P(s) = \frac{N(s)}{D(s)}$ will be expressed as a sum of simple fractions whose inverse transforms are easy to obtain. For this purpose, the Heaviside method will be used. Based on the roots of $D(s)$, also called poles, the following case studies will be considered:
  - Simple real poles.
  - Multiple real poles.
  - Complex-conjugated poles.

3.6.2 Simple Real Poles, $p_1 \neq p_2 \neq \cdots \neq p_m$

Let be

$$P(s) = \frac{N(s)}{D(s)} = \frac{N(s)}{(s-p_1)\ldots(s-p_m)} = \frac{R_1}{s-p_1} + \frac{R_2}{s-p_2} + \cdots + \frac{R_m}{s-p_m} \tag{3.26}$$

where R_k are called residuals and p_k are the roots of $D(s)$, so for $k = 1, \ldots, m$:

$$R_k = (s-p_k)\, P(s)\big|_{s=p_k} = \frac{N(p_k)}{\prod\limits_{\substack{i=1\\ i\neq k}}^{m}(p_k - p_i)} \tag{3.27}$$

Applying the inverse transform:

$$p(t) = \mathscr{L}^{-1}[P(s)] = R_1\, \mathrm{e}^{p_1 t} + \cdots + R_m\, \mathrm{e}^{p_m t} \tag{3.28}$$

Example. Simple real poles

$$P(s) = \frac{s+4}{s^2+4s+3} = \frac{s+4}{(s+1)(s+3)} = \frac{R_1}{s+1} + \frac{R_2}{s+3}$$

Poles: $p_1 = -1$; $p_2 = -3$

Solution Calculation of the residuals:

$$R_1 = (s+1)\, P(s)\big|_{s=-1} = \left.\frac{s+4}{s+3}\right|_{s=-1} = \frac{-1+4}{-1+3} = \frac{3}{2}$$

$$R_2 = (s+3)\, P(s)\big|_{s=-3} = \left.\frac{s+4}{s+1}\right|_{s=-3} = \frac{-3+4}{-3+1} = \frac{-1}{2}$$

The following result is obtained:

$$P(s) = \frac{3/2}{s+1} + \frac{-1/2}{s+3}$$

The inverse transform of each fraction can be calculated according to the relationships presented in Table 3.1:

$$\mathcal{L}^{-1}\left[\frac{1}{s+a}\right] = \mathrm{e}^{-at}$$

Time function:

$$p(t) = \mathcal{L}^{-1}[P(s)] = \mathcal{L}^{-1}\left[\frac{3/2}{s+1} + \frac{-1/2}{s+3}\right] = \frac{3}{2}\cdot\mathrm{e}^{-t} - \frac{1}{2}\cdot\mathrm{e}^{-3t}$$

3.6.3 *Multiple Real Pole*

Let be

$$P(s) = \frac{N(s)}{D(s)} = \frac{N(s)}{(s-p_1)^m} = \frac{R_1}{s-p_1} + \cdots + \frac{R_m}{(s-p_1)^m} \tag{3.29}$$

the residuals are calculated as:

$$\begin{aligned} R_m &= (s-p_1)^m\, P(s)\big|_{s=p_1} \\ R_{m-1} &= \frac{1}{1!}\frac{d[(s-p_1)^m\, P(s)]}{ds}\bigg|_{s=p_1} \\ &\vdots \\ R_1 &= \frac{1}{(m-1)!}\frac{d^{m-1}[(s-p_1)^m\, P(s)]}{ds^{m-1}}\bigg|_{s=p_1} \end{aligned} \tag{3.30}$$

Yielding the following result:

$$p(t) = \mathcal{L}^{-1}[P(s)] = \left(R_1 + R_2 t + \cdots + R_m t^{m-1}\right)\mathrm{e}^{p_1 t} \tag{3.31}$$

Example. Double real pole

$$P(s) = \frac{s+2}{(s+1)^2} = \frac{R_1}{s+1} + \frac{R_2}{(s+1)^2}$$

Pole: $p_1 = -1$ (double)

Solution Calculation of the residuals:

$$R_2 = (s+1)^2\, P(s)\big|_{s=-1} = (s+2)|_{s=-1} = -1+2 = 1$$

$$R_1 = \left.\frac{d[(s+1)^2\, P(s)]}{ds}\right|_{s=-1} = \left.\frac{d[s+2]}{ds}\right|_{s=-1} = 1$$

Being obtained the following expression:

$$P(s) = \frac{1}{s+1} + \frac{1}{(s+1)^2}$$

The inverse transform of each fraction can be derived according to the relationships in Table 3.1:

$$\mathscr{L}^{-1}\left[\frac{1}{s+a}\right] = \mathrm{e}^{-at}$$

$$\mathscr{L}^{-1}\left[\frac{r!}{(s+a)^{r+1}}\right] = t^r\,\mathrm{e}^{-at}$$

Time function:

$$p(t) = \mathscr{L}^{-1}[P(s)] = \mathscr{L}^{-1}\left[\frac{1}{s+1} + \frac{1}{(s+1)^2}\right] = (1+t)\,\mathrm{e}^{-t}$$

3.6.4 Complex-Conjugated Pole

Let be

$$P(s) = \frac{N(s)}{D(s)} = \frac{N(s)}{[s-(\alpha+j\beta)][s-(\alpha-j\beta)]} = \frac{R\angle\psi}{s-(\alpha+j\beta)} + \frac{R\angle-\psi}{s-(\alpha-j\beta)} \tag{3.32}$$

with

$$p_1, p_2 = \alpha \pm j\beta \ , \ \ (\beta \geq 0) \tag{3.33}$$

Then $R\angle\psi$ is a complex number that is calculated as follows:

$$R\angle\psi = [s-(\alpha+j\beta)]\, P(s)|_{s=\alpha+j\beta} \tag{3.34}$$

Obtaining the inverse transform as follows:

$$p(t) = \mathscr{L}^{-1}[P(s)] = 2\, R\, \mathrm{e}^{\alpha\, t}\cos(\beta\, t + \psi) \tag{3.35}$$

Example. Complex-conjugated pole

$$P(s)=\frac{s+4}{s^2+2s+5}=\frac{s+4}{[s-(-1+j2)][s-(-1-j2)]}$$
$$=\frac{R\angle\psi}{s-(-1+j2)}+\frac{R\angle-\psi}{s-(-1-j2)}$$

Poles: $p_1,\ p_2=\alpha\pm j\beta=-1\pm j2$

Solution Calculation of the residuals:

$$R\angle\psi=\left.\frac{[s-(-1+j2)]\,(s+4)}{[s-(-1+j2)][s-(-1-j2)]}\right|_{s=-1+j2}=\left.\frac{s+4}{s-(-1-j2)}\right|_{s=-1+j2}$$
$$=\frac{3+j2}{j4}=0{,}9\angle-56{,}31^\circ$$

Resulting in the following expression:

$$P(s)=\frac{0{,}9\angle-56{,}31^\circ}{s-(-1+j2)}+\frac{0{,}9\angle56{,}31^\circ}{s-(-1-j2)}$$

The inverse transform of $P(s)$ can be obtained according to the relationships in Table 3.1:

$$\mathscr{L}^{-1}\left[\frac{R\angle\psi}{[s-(\alpha+j\beta)]}+\frac{R\angle-\psi}{[s-(\alpha-j\beta)]}\right]=2Re^{\alpha t}\cos(\beta t+\psi)$$

Time function:

$$p(t)=\mathscr{L}^{-1}\left[\frac{0{,}9\angle-56{,}31^\circ}{s-(-1+j2)}+\frac{0{,}9\angle56{,}31^\circ}{s-(-1-j2)}\right]=1{,}8\,\mathrm{e}^{-t}\cos(2t-56{,}31^\circ)$$

3.7 Solved Problems

3.7.1 First Order Circuit Under DC Supply

In the circuit represented in Fig. 3.20, calculate $I_L(s)$, transform from $i_L(t)$. Then, obtain the time expression for $t>0$.

Solution. First, in order transform the inductor to the Laplace domain, the current at $t=0^-$ must be known. Since for $t<0$ the switch is open, there is no current flow ($i_L(0^-)=0\,\mathrm{A}$).

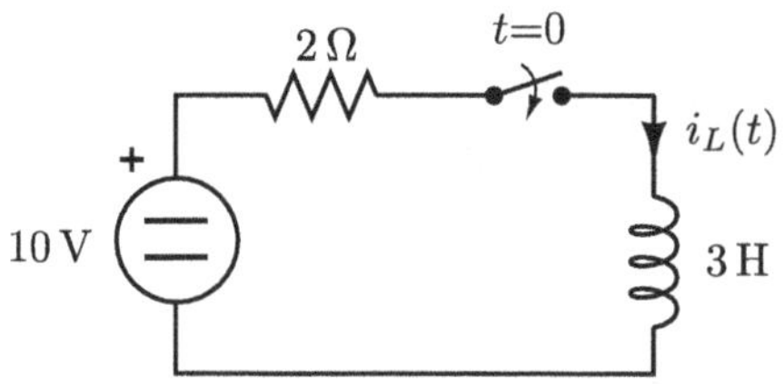

Fig. 3.20 Circuit considered in the problem under study

The circuit for $t > 0$ is that shown in Fig. 3.21 in the Laplace domain. From the circuit in Fig. 3.21 the current $I_L(s)$ is calculated:

$$I_L(s) = \frac{10/s}{2+3s} = \frac{10/s}{3(s+\frac{2}{3})} = \frac{10/3}{s(s+\frac{2}{3})}$$

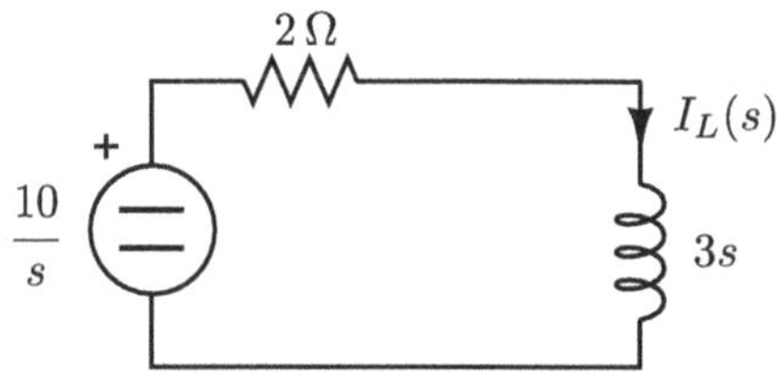

Fig. 3.21 Circuit in the Laplace domain

Using the partial fraction expansion results in:

$$I_L(s) = \frac{10/3}{s(s+\frac{2}{3})} = \frac{A}{s} + \frac{B}{(s+\frac{2}{3})}$$

The coefficients A and B are calculated as follows:

$$A = \left.\frac{10/3}{(s+\frac{2}{3})}\right|_{s=0} = \frac{10/3}{2/3} = 5$$

$$B = \left.\frac{10/3}{s}\right|_{s=\frac{-2}{3}} = \frac{10/3}{-2/3} = -5$$

So:

$$I_L(s) = \frac{5}{s} - \frac{5}{(s+\frac{2}{3})}\ \text{A}$$

Finally, the inverse Laplace transform is taken:

$$i_L(t) = \mathscr{L}^{-1}\left[I_L(s)\right] = 5 - 5\cdot e^{-2t/3}\ \text{A}$$

3.7.2 First Order Circuit Under DC Supply

In the circuit represented in Fig. 3.22, it is known that $u_C(0^-) = 0\,\text{V}$. Calculate $U_C(s)$, transform of $u_C(t)$. Then obtain the time expression, $u_C(t)$, for $t > 0$.

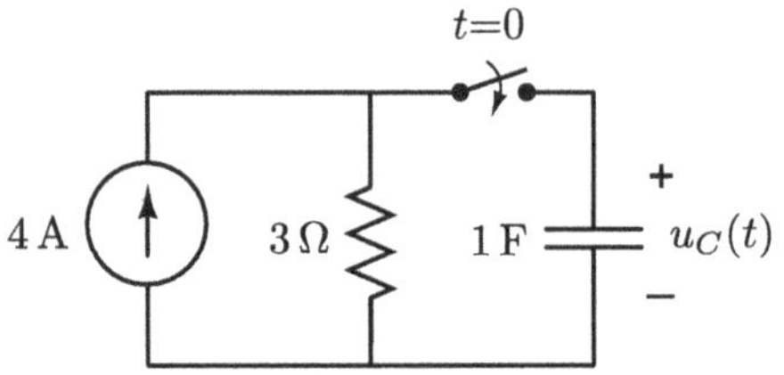

Fig. 3.22 Circuit considered in the problem under study

Solution. The circuit for $t > 0$ is that shown in Fig. 3.23 in the Laplace domain.

From the circuit in Fig. 3.23, the current $I_C(s)$ is calculated using the concept of intensity divider:

$$I_C(s) = \frac{4}{s} \cdot \frac{s}{s + \frac{1}{3}} = \frac{4}{s + \frac{1}{3}}$$

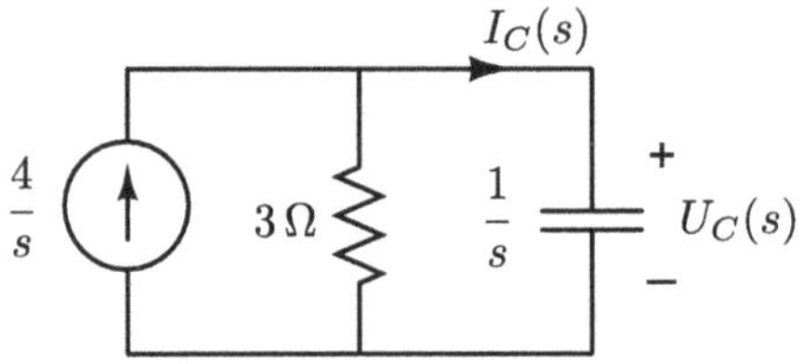

Fig. 3.23 Circuit in the Laplace domain

Therefore, the voltage $U_C(s)$ is the following:

$$U_C(s) = \frac{1}{s} \cdot I(s) = \frac{1}{s} \cdot \frac{4}{s + \frac{1}{3}} = \frac{4}{s(s + \frac{1}{3})}$$

Using the partial fraction expansion results in:

$$U_C(s) = \frac{4}{s(s + \frac{1}{3})} = \frac{A}{s} + \frac{B}{(s + \frac{1}{3})}$$

The coefficients A and B are calculated as follows:

$$A = \left. \frac{4}{(s + \frac{1}{3})} \right|_{s=0} = \frac{4}{1/3} = 12$$

$$B = \left. \frac{4}{s} \right|_{s=\frac{-1}{3}} = \frac{4}{-1/3} = -12$$

So:

$$U_C(s) = \frac{12}{s} - \frac{12}{(s + \frac{1}{3})} \text{ V}$$

Finally, the inverse Laplace transform is taken:

$$u_C(t) = \mathscr{L}^{-1}\left[U_C(s)\right] = 12 - 12 \cdot \mathrm{e}^{-t/3} \text{ V}$$

3.7.3 First Order Circuit Under DC Supply

The circuit represented in Fig. 3.24 is in steady state when at $t = 0$ the switch opens. Calculate $I(s)$, transform of $i(t)$. Then, obtain the time expression, $i(t)$, for $t > 0$.

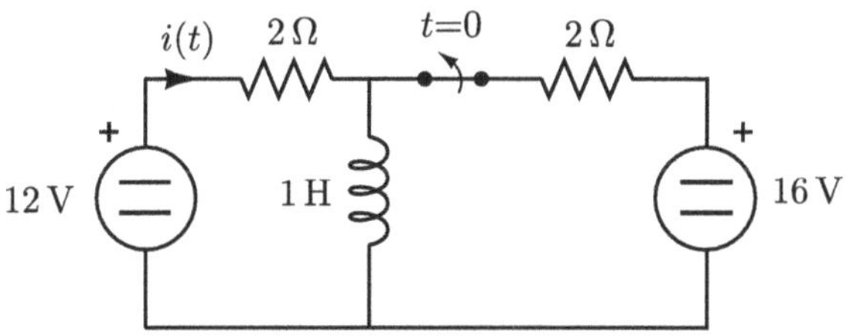

Fig. 3.24 Circuit considered in the problem under study

Solution. First, the inductor current will be obtained before the switch opens, that is, at instant $t = 0^-$. For this purpose, it will be considered that the circuit is in steady-state regime before the switch is opened. Since only dc-type excitation sources are present, the dc circuit shown in Fig. 3.25 will be solved, where the inductor has been replaced by a short circuit.

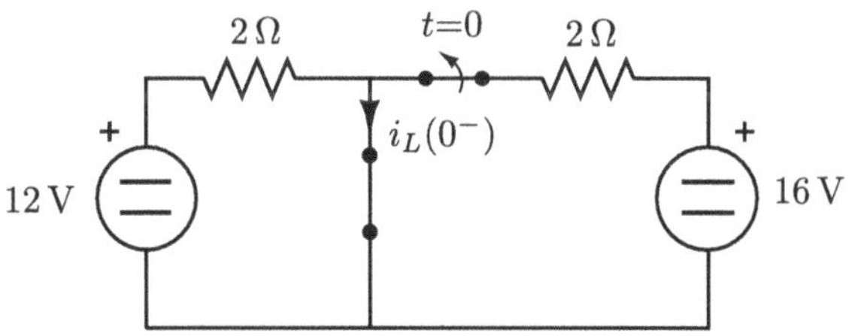

Fig. 3.25 Circuit at $t = 0^-$

From the circuit in Fig. 3.25 the current through the inductor is obtained at $t = 0^-$:

$$i_L(0^-) = \frac{12}{2} + \frac{16}{2} = 6 + 8 = 14 \text{ A}$$

The circuit for $t > 0$ is that shown in Fig. 3.26 in the Laplace domain, according to which:

$$I(s) = \frac{\frac{12}{s} + 14}{s+2} = \frac{14s+12}{s(s+2)}$$

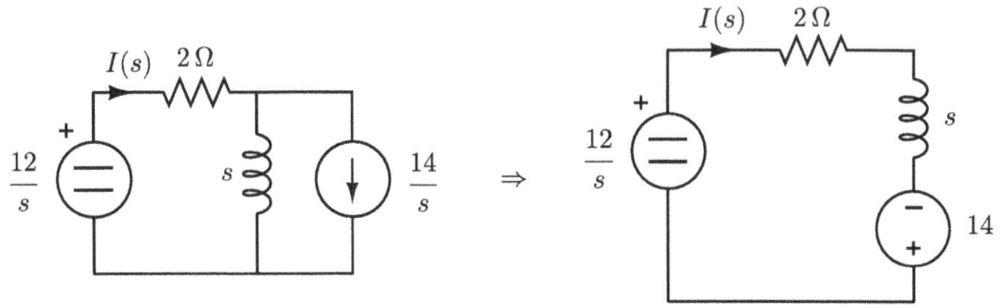

Fig. 3.26 Circuit in the Laplace domain

Using the partial fraction expansion results in:

$$I(s) = \frac{14s+12}{s(s+2)} = \frac{A}{s} + \frac{B}{(s+2)}$$

The coefficients A and B are calculated as follows:

$$A = \left.\frac{14s+12}{(s+2)}\right|_{s=0} = \frac{12}{2} = 6$$

$$B = \left.\frac{14s+12}{s}\right|_{s=-2} = \frac{12-28}{-2} = 8$$

So:

$$I(s) = \frac{6}{s} + \frac{8}{(s+2)}\ \text{A}$$

Finally, the inverse Laplace transform is taken:

$$i(t) = \mathscr{L}^{-1}\left[I(s)\right] = 6 + 8 \cdot \mathrm{e}^{-2t}\ \text{A}$$

3.7.4 First Order Circuit Under DC Supply

The circuit represented in Fig. 3.27 is in steady state when, at $t = 0$, the switch k1 is closed and k2 is opened. Calculate $I(s)$, transform of $i(t)$. Then, obtain the time expression, $i(t)$, for $t > 0$.

Solution. First, the capacitor voltage will be obtained before the switches change their position, that is, at instant $t = 0^-$. For this purpose, it will be considered that the

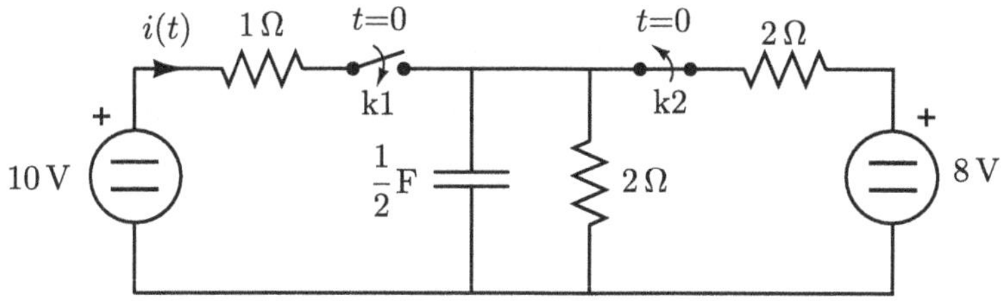

Fig. 3.27 Circuit considered in the problem under study

circuit is in steady-state regime before switch k1 closes and k2 opens. Since, in this situation, only a dc-type source of excitation is present, the dc circuit steady-state regime shown in Fig. 3.28 will be solved, where the capacitor has been replaced by an open circuit.

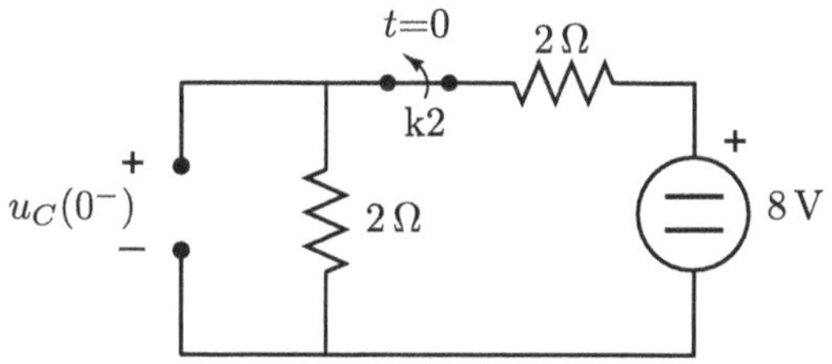

Fig. 3.28 Circuit at $t = 0^-$

From the circuit in Fig. 3.28 the capacitor voltage at $t = 0^-$ is obtained as:

$$u_C(0^-) = 8 \cdot \frac{2}{2+2} = 4\,\text{V}$$

The circuit for $t > 0$ is that shown in Fig. 3.29 in the Laplace domain.

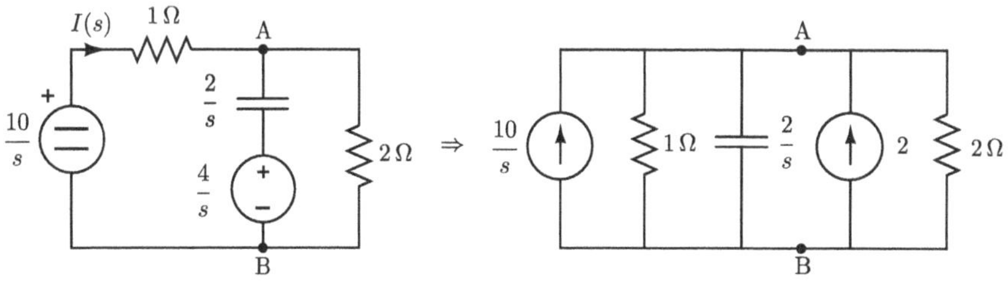

Fig. 3.29 Circuit in the Laplace domain

From the circuit in Fig. 3.29, and applying the node analysis method, the voltage between nodes A and B is obtained:

$$\left(1+\frac{s}{2}+\frac{1}{2}\right)\cdot U_{AB}(s)=\frac{10}{s}+2\ \rightarrow\ U_{AB}(s)=\frac{2\cdot(2s+10)}{s(s+3)}$$

On the other hand:

$$U_{AB}(s)=\frac{10}{s}-1\cdot I(s)$$

Solving and substituting values results in:

$$I(s)=\frac{10}{s}-U_{AB}(s)=\frac{6s+10}{s(s+3)}$$

Using the partial fraction expansion:

$$I(s)=\frac{6s+10}{s(s+3)}=\frac{A}{s}+\frac{B}{(s+3)}$$

The coefficients A and B are calculated as follows:

$$A=\left.\frac{6s+10}{(s+3)}\right|_{s=0}=\frac{10}{3}$$

$$B=\left.\frac{6s+10}{s}\right|_{s=-3}=\frac{-18+10}{-3}=\frac{8}{3}$$

So:

$$I(s)=\frac{10/3}{s}+\frac{8/3}{(s+3)}\ \mathrm{A}$$

Finally, the inverse Laplace transform is taken:

$$i(t)=\mathscr{L}^{-1}\left[I(s)\right]=\frac{10}{3}+\frac{8}{3}\cdot\mathrm{e}^{-3t}\ \mathrm{A}$$

3.7.5 First Order Circuit Under AC Supply

The circuit represented in Fig. 3.30 is in steady state when, at t=0, the switches change position. Given that $e(t)$=$100\sqrt{2}\sin(50t)$ V, calculate $I_L(s)$, transform of $i_L(t)$. Then, obtain the time expression, $i_L(t)$, for $t > 0$.

Solution. First, $i_L(0^-)$ must be calculated. For this purpose, the steady-state regime prior to the position change of the switches is analyzed. The circuit corresponding to the steady-state dc regime prior to the position change of the switches is shown in Fig. 3.31, where the inductor has been replaced by a short circuit.

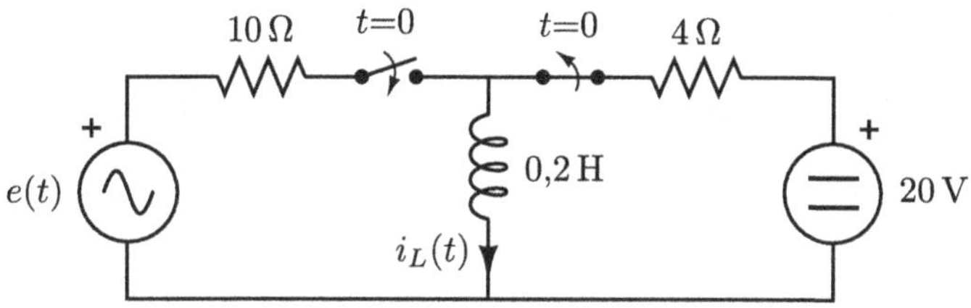

Fig. 3.30 Circuit considered in the problem under study

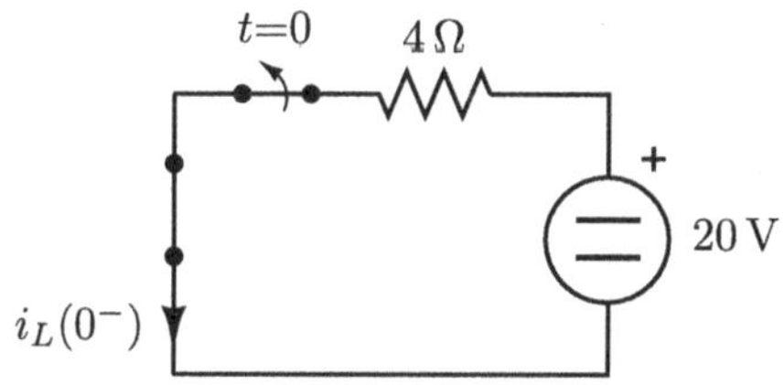

Fig. 3.31 Circuit at $t = 0^-$

From the circuit in Fig. 3.31:

$$i_L(0^-) = \frac{20}{4} = 5\,\mathrm{A}$$

The circuit for $t > 0$ is that shown in Fig. 3.32 in the Laplace domain, where

$$E(s) = \mathscr{L}[e(t)] = \mathscr{L}\left[100\sqrt{2}\sin(50t)\right] = \frac{100\sqrt{2}\cdot 50}{s^2+50^2} = \frac{5\,000\sqrt{2}}{s^2+50^2}$$

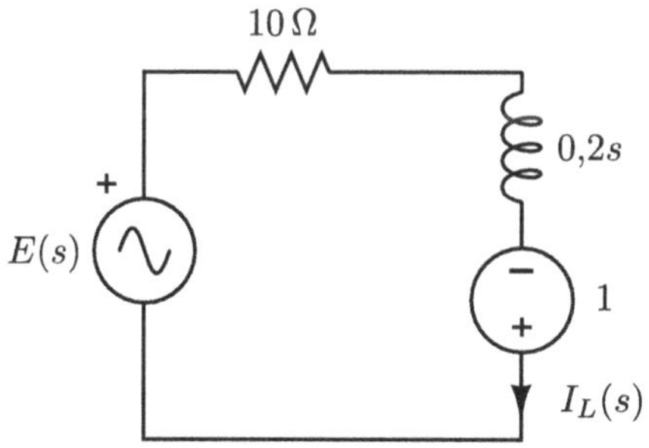

Fig. 3.32 Circuit in the Laplace domain

From the circuit in Fig. 3.32, the current $I_L(s)$ is calculated:

$$I_L(s) = \frac{\frac{5\,000\sqrt{2}}{s^2+50^2}+1}{0{,}2s+10} = \frac{\frac{5\,000\sqrt{2}+s^2+50^2}{s^2+50^2}}{0{,}2(s+50)} = \frac{\frac{1}{0{,}2}\cdot(5\,000\sqrt{2}+s^2+50^2)}{(s^2+50^2)(s+50)}$$
$$= \frac{\frac{1}{0{,}2}\cdot(5\,000\sqrt{2}+s^2+50^2)}{(s-50j)(s+50j)(s+50)}$$

Using the partial fraction expansion yields:

$$I_L(s) = \frac{\frac{1}{0{,}2}\cdot(5\,000\sqrt{2}+s^2+50^2)}{(s-50j)(s+50j)(s+50)} = \frac{\overline{A}}{(s-50j)} + \frac{\overline{A}^*}{(s+50j)} + \frac{B}{(s+50)}$$

The coefficients $\overline{A}$ and B are calculated as follows:

$$\overline{A} = \left.\frac{\frac{1}{0{,}2}\cdot(5\,000\sqrt{2}+s^2+50^2)}{(s+50j)(s+50)}\right|_{s=50j}$$
$$= \frac{\frac{1}{0{,}2}\cdot(5\,000\sqrt{2}+(50j)^2+50^2)}{(50j+50j)(50j+50)} = 5\angle-135^\circ$$

$$B = \left.\frac{\frac{1}{0{,}2}\cdot(5\,000\sqrt{2}+s^2+50^2)}{(s+50j)(s-50j)}\right|_{s=-50}$$
$$= \frac{\frac{1}{0{,}2}\cdot(5\,000\sqrt{2}+(-50)^2+50^2)}{(-50+50j)(-50-50j)} \approx 12{,}07$$

So:

$$I_L(s) = \frac{5\angle-135^\circ}{(s-50j)} + \frac{5\angle135^\circ}{(s+50j)} + \frac{12{,}07}{(s+50)}\,\text{A}$$

Finally, the inverse Laplace transform is taken:

$$i_L(t) = \mathscr{L}^{-1}\left[I_L(s)\right] = 2\cdot5\cdot\cos(50t-135^\circ) + 12{,}07\cdot\mathrm{e}^{-50t}\,\text{A}$$

3.7.6 First Order Circuit Under DC and AC Supply

In the circuit represented in Fig. 3.33, both switches are closed at $t = 0$. Given that $u_C(0^-) = 5\,\text{V}$ and that $e(t) = 6\cos(10t)\,\text{V}$, calculate $U_C(s)$, transform of $u_C(t)$. Then, obtain the time expression, $u_C(t)$, for $t > 0$.

Solution. The circuit for $t > 0$ is that shown in Fig. 3.34 in the Laplace domain, where

$$E(s) = \mathscr{L}\left[e(t)\right] = \mathscr{L}\left[6\cos(10t)\right] = \frac{6s}{s^2+10^2}$$

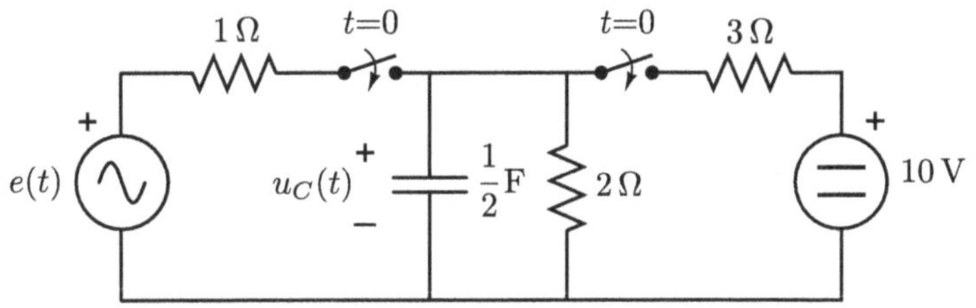

Fig. 3.33 Circuit considered in the problem under study

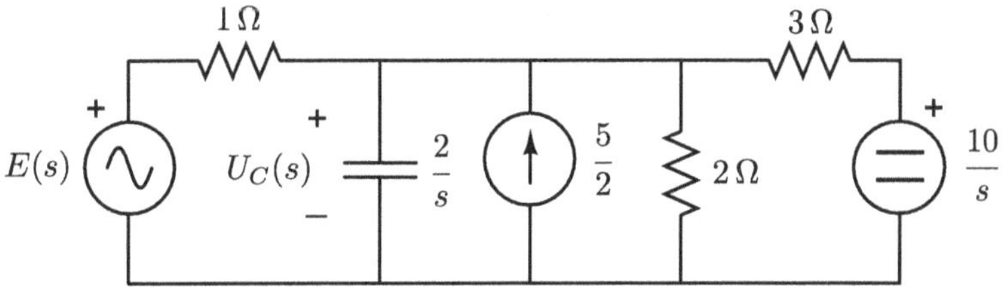

Fig. 3.34 Circuit in the Laplace domain

In this case, for convenience, the initial condition of the capacitor has been modeled as a current source.

From the circuit in Fig. 3.34, replacing the real voltage sources with real current sources, and applying the node analysis method, the following result is obtained:

$$\left(1+\frac{s}{2}+\frac{1}{2}+\frac{1}{3}\right)\cdot U_C(s)=\frac{E(s)}{1}+\frac{5}{2}+\frac{10}{3s}$$

Rearranging terms:

$$U_C(s)=\frac{6s\cdot E(s)+20+15s}{s(3s+11)}=\frac{6s\cdot E(s)+20+15s}{3s(s+\frac{11}{3})}$$

Substituting the value of $E(s)$:

$$U_C(s)=\frac{6s\cdot\frac{6s}{s^2+10^2}+20+15s}{3s(s+\frac{11}{3})}=\frac{15s^3+56s^2+1\,500s+2\,000}{3s(s^2+10^2)(s+\frac{11}{3})}$$

Using the partial fraction expansion results in:

$$U_C(s)=\frac{15s^3+56s^2+1\,500s+2\,000}{3s(s-10j)(s+10j)(s+\frac{11}{3})}=\frac{A}{s}+\frac{\overline{B}}{(s-10j)}+\frac{\overline{B}^*}{(s+10j)}+\frac{C}{(s+\frac{11}{3})}$$

The coefficients A, $\overline{B}$ and C are calculated as follows:

$$A = \left.\frac{15s^3 + 56s^2 + 1\,500s + 2\,000}{3(s-10j)(s+10j)(s+\frac{11}{3})}\right|_{s=0} = \frac{2\,000}{1\,100} = \frac{20}{11}$$

$$\overline{B} = \left.\frac{15s^3 + 56s^2 + 1\,500s + 2\,000}{3s(s+10j)(s+\frac{11}{3})}\right|_{s=10j}$$
$$= \frac{15(10j)^3 + 56(10j)^2 + 1\,500(10j) + 2\,000}{3(10j)(10j+10j)(10j+\frac{11}{3})} \approx 0{,}56\angle - 69{,}86^\circ$$

$$C = \left.\frac{15s^3 + 56s^2 + 1\,500s + 2\,000}{3s(s-10j)(s+10j)}\right|_{s=-11/3}$$
$$= \frac{15(-11/3)^3 + 56(-11/3)^2 + 1\,500(-11/3) + 2\,000}{3(-11/3)(-11/3-10j)(-11/3+10j)} \approx 2{,}79$$

So:

$$U_C(s) = \frac{20/11}{s} + \frac{0{,}56\angle - 69{,}86^\circ}{(s-10j)} + \frac{0{,}56\angle 69{,}86^\circ}{(s+10j)} + \frac{2{,}79}{(s+\frac{11}{3})}\ \text{V}$$

Finally, the inverse Laplace transform is taken:

$$u_C(t) = \mathscr{L}^{-1}\left[U_C(s)\right] = \frac{20}{11} + 2\cdot 0{,}56 \cdot \cos(10t - 69{,}86^\circ) + 2{,}79 \cdot \mathrm{e}^{-11t/3}\ \text{V}$$

3.7.7 Overdamped Second Order Circuit Without Supply

In the circuit represented in Fig. 3.35, given that $i_L(0^-)$=3 A and $u_C(0^-)$=2 V, calculate $I_L(s)$, transform of $i_L(t)$. Then, obtain the time expression, $i_L(t)$, for $t > 0$.

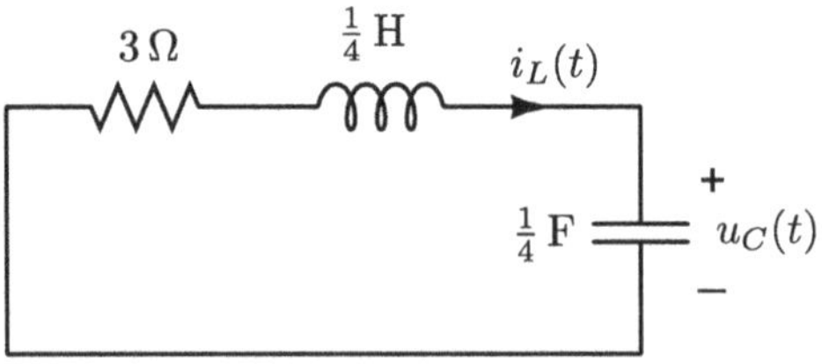

Fig. 3.35 Circuit considered in the problem under study

Solution. The circuit for $t > 0$ is that shown in Fig. 3.36 in the Laplace domain.
From the circuit in Fig. 3.36, the current $I_L(s)$ is calculated as:

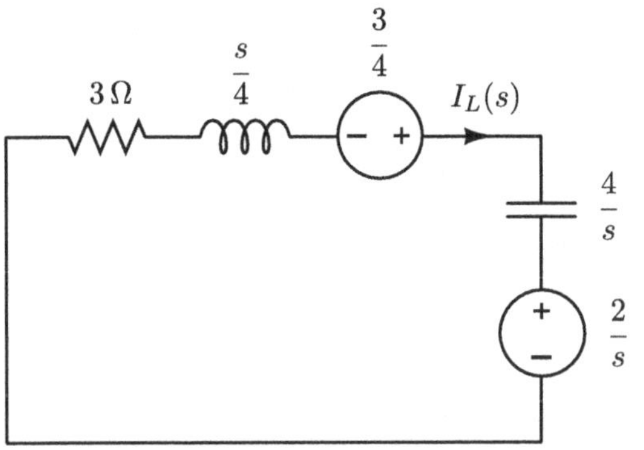

Fig. 3.36 Circuit in the Laplace domain

$$I_L(s) = \frac{\frac{3}{4} - \frac{2}{s}}{3 + \frac{s}{4} + \frac{4}{s}} = \frac{\frac{3s-8}{4s}}{\frac{s^2+12s+16}{4s}} = \frac{3s-8}{s^2+12s+16} \approx \frac{3s-8}{(s+1{,}53)(s+10{,}47)}$$

Using the partial fraction expansion results in:

$$I_L(s) = \frac{3s-8}{(s+1{,}53)(s+10{,}47)} = \frac{A}{(s+1{,}53)} + \frac{B}{(s+10{,}47)}$$

The coefficients A and B are calculated as follows:

$$A = \left.\frac{3s-8}{(s+10{,}47)}\right|_{s=-1{,}53} = \frac{3\cdot(-1{,}53)-8}{(-1{,}53+10{,}47)} \approx -1{,}41$$

$$B = \left.\frac{3s-8}{(s+1{,}53)}\right|_{s=-10{,}47} = \frac{3\cdot(-10{,}47)-8}{(-10{,}47+1{,}53)} \approx 4{,}41$$

So:

$$I_L(s) = \frac{-1{,}41}{(s+1{,}53)} + \frac{4{,}41}{(s+10{,}47)}\ \text{A}$$

Finally, the inverse Laplace transform is taken:

$$i_L(t) = \mathscr{L}^{-1}\left[I_L(s)\right] = -1{,}41\cdot \mathrm{e}^{-1{,}53t} + 4{,}41\cdot \mathrm{e}^{-10{,}47t}\ \text{A}$$

3.7.8 Underdamped Second Order Circuit Under AC Supply

In the circuit represented in Fig. 3.37, the capacitor is discharged before the switch closes. Given that $e(t) = 30\cos(5t)$ V, calculate $I(s)$, transform of $i(t)$. Then, obtain the time exprcssion, $i(t)$, for $t > 0$.

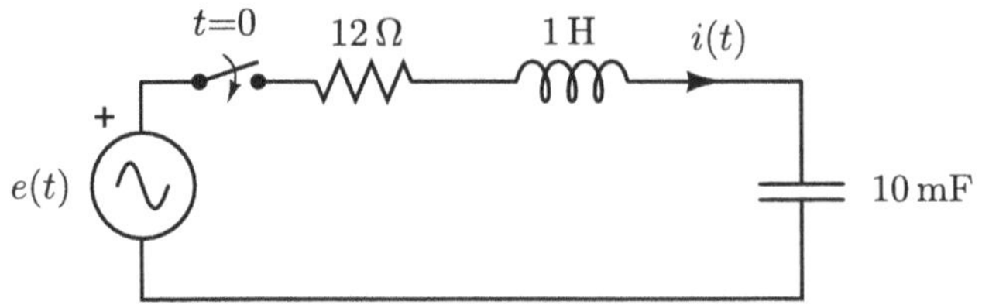

Fig. 3.37 Circuit considered in the problem under study

Solution. Before the switch is closed, the inductor current is zero ($i_L(0^-)$=0 A). On the other hand, according to the problem statement, the capacitor is discharged before closing the switch, therefore $u_C(0^-)$=0 V.

The circuit for $t > 0$ is that shown in Fig. 3.38 in the Laplace domain, where:

$$E(s) = \mathscr{L}\left[e(t)\right] = \mathscr{L}\left[30\cos(5t)\right] = \frac{30s}{s^2+5^2}$$

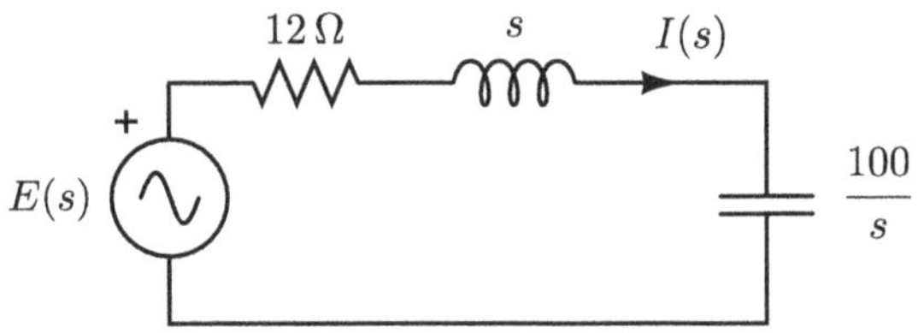

Fig. 3.38 Circuit in the Laplace domain

From the circuit in Fig. 3.38, the current $I(s)$ is calculated as:

$$I(s) = \frac{E(s)}{12+s+\frac{100}{s}} = \frac{\frac{30s}{s^2+5^2}}{12+s+\frac{100}{s}} = \frac{30s^2}{(s^2+5^2)(s^2+12s+100)}$$
$$= \frac{30s^2}{(s-5j)(s+5j)(s+6-8j)(s+6+8j)}$$

Using the partial fraction expansion results in:

$$I(s) = \frac{30s^2}{(s-5j)(s+5j)(s+6-8j)(s+6+8j)}$$
$$= \frac{\overline{A}}{(s-5j)} + \frac{\overline{A}^*}{(s+5j)} + \frac{\overline{B}}{(s+6-8j)} + \frac{\overline{B}^*}{(s+6+8j)}$$

The coefficients $\overline{A}$ and $\overline{B}$ are calculated as follows:

$$\overline{A} = \left.\frac{30s^2}{(s+5j)(s+6-8j)(s+6+8j)}\right|_{s=5j}$$
$$= \frac{30(5j)^2}{(5j+5j)(5j+6-8j)(5j+6+8j)} = \frac{-750}{-600+750j} \approx 0{,}78\angle 51{,}34°$$

$$\overline{B} = \left.\frac{30s^2}{(s-5j)(s+5j)(s+6+8j)}\right|_{s=-6+8j}$$
$$= \frac{30(-6+8j)^2}{(-6+8j-5j)(-6+8j+5j)(-6+8j+6+8j)} \approx 1{,}95\angle -104{,}47°$$

So:

$$I(s) = \frac{0{,}78\angle 51{,}34°}{(s-5j)} + \frac{0{,}78\angle -51{,}34°}{(s+5j)} + \frac{1{,}95\angle -104{,}47°}{(s+6-8j)} + \frac{1{,}95\angle 104{,}47°}{(s+6+8j)} \text{ A}$$

Finally, the inverse Laplace transform is taken:

$$i(t) = \mathscr{L}^{-1}[I(s)] = 2\cdot 0{,}78\cdot\cos(5t+51{,}34°) + 2\cdot 1{,}95\cdot e^{-6t}\cdot\cos(8t-104{,}47°) \text{ A}$$

3.7.9 Overdamped Second Order Circuit Under AC Supply

The circuit represented in Fig. 3.39 is in steady-state before the switches change their position. Given that $\frac{1}{2}$ F capacitor is initially discharged, $e_1(t) = 50\sin(10t)$ V and $e_2(t) = 25\sin(20t)$ V, calculate $I_L(s)$, transform of $i_L(t)$. Then, obtain the time expression, $i_L(t)$, for $t > 0$.

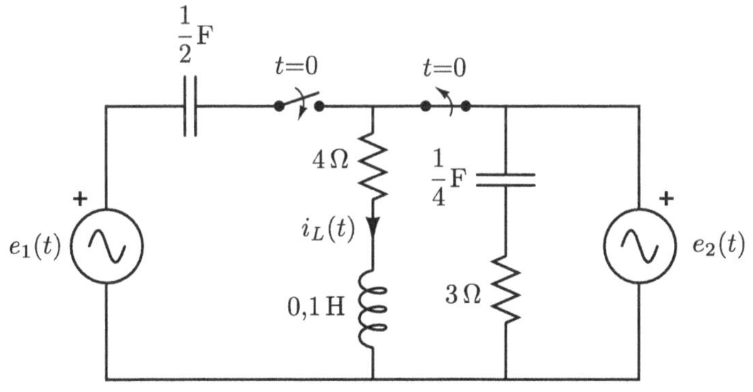

Fig. 3.39 Circuit considered in the problem under study

Solution. First, the inductor current will be obtained before the switch closes, that is, at instant $t = 0^-$. For this purpose, it will be considered that the circuit is in steady-state regime before the switch closes. Since only ac-type excitation sources are present, the circuit in steady-state ac regime will be solved in the phasor domain, as shown in Fig. 3.40.

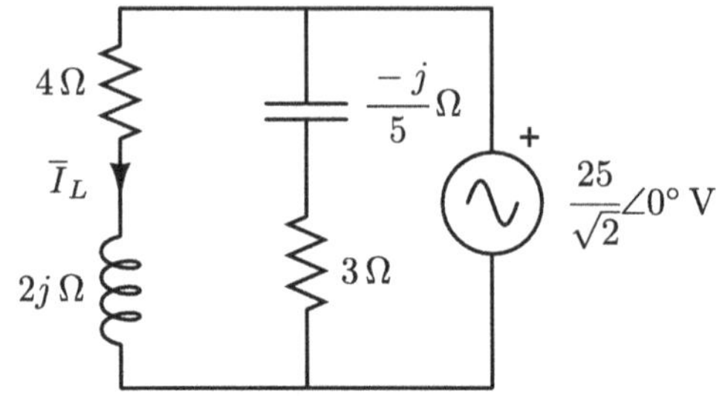

Fig. 3.40 Circuit in the phasor domain

From the circuit in Fig. 3.40:

$$\overline{I}_L = \frac{\frac{25}{\sqrt{2}}\angle 0°}{4+2j} \approx 3{,}95\angle - 26{,}57°\text{ A}$$

The corresponding time function is the following:

$$i_L(t) = \sqrt{2} \cdot 3{,}95 \sin(20t - 26{,}57°)\text{ A}$$

For $t = 0^-$, the value of $i_L(t)$ is:

$$i_L(0^-) = \sqrt{2} \cdot 3{,}95 \sin(-26{,}57°) \approx -2{,}5\text{ A}$$

The circuit for $t > 0$ is that shown in Fig. 3.41 in the Laplace domain, where:

$$E_1(s) = \mathscr{L}[e_1(t)] = \mathscr{L}[50\sin(10t)] = \frac{50 \cdot 10}{s^2 + 10^2}$$

From the circuit in Fig. 3.41, the current $I_L(s)$ is calculated as:

$$I_L(s) = \frac{E_1(s) - 0{,}25}{\frac{2}{s} + 4 + 0{,}1s} = \frac{\frac{50 \cdot 10}{s^2+10^2} - 0{,}25}{\frac{2}{s} + 4 + 0{,}1s} = \frac{-2{,}5s^3 + 4\,750s}{(s^2+10^2)(s^2+40s+20)}$$

Using the partial fraction expansion results in:

$$\begin{aligned} I_L(s) &= \frac{-2{,}5s^3 + 4\,750s}{(s-10j)(s+10j)(s+39{,}5)(s+0{,}5)} \\ &= \frac{\overline{A}}{(s-10j)} + \frac{\overline{A}^*}{(s+10j)} + \frac{B}{(s+39{,}5)} + \frac{C}{(s+0{,}5)} \end{aligned}$$

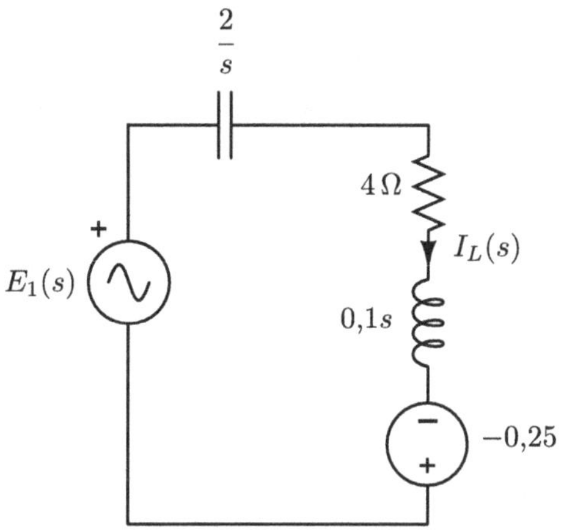

Fig. 3.41 Circuit in the Laplace domain

The coefficients $\overline{A}$, B and C are calculated as follows:

$$\overline{A} = \left.\frac{-2{,}5s^3 + 4\,750s}{(s+10j)(s+39{,}5)(s+0{,}5)}\right|_{s=10j}$$

$$= \frac{-2{,}5(10j)^3 + 4\,750(10j)}{(10j+10j)(10j+39{,}5)(10j+0{,}5)} \approx \frac{50\,000\angle 90^\circ}{8\,159\angle -168{,}66^\circ} \approx 6{,}13\angle -101{,}34^\circ$$

$$B = \left.\frac{-2{,}5s^3 + 4\,750s}{(s-10j)(s+10j)(s+0{,}5)}\right|_{s=-39{,}5}$$

$$= \frac{-2{,}5(-39{,}5)^3 + 4\,750(-39{,}5)}{(-39{,}5-10j)(-39{,}5+10j)(-39{,}5+0{,}5)} \approx 0{,}52$$

$$C = \left.\frac{-2{,}5s^3 + 4\,750s}{(s-10j)(s+10j)(s+39{,}5)}\right|_{s=-0{,}5}$$

$$= \frac{-2{,}5(-0{,}5)^3 + 4\,750(-0{,}5)}{(-0{,}5-10j)(-0{,}5+10j)(-0{,}5+39{,}5)} \approx -0{,}61$$

So:

$$I_L(s) = \frac{6{,}13\angle -101{,}34^\circ}{(s-10j)} + \frac{6{,}13\angle 101{,}34^\circ}{(s+10j)} + \frac{0{,}52}{(s+39{,}5)} + \frac{-0{,}61}{(s+0{,}5)}\ \text{A}$$

Finally, the inverse Laplace transform is taken:

$$i_L(t) = \mathscr{L}^{-1}\left[I_L(s)\right] = 2\cdot 6{,}13\cdot\cos(10t - 101{,}34^\circ) + 0{,}52\cdot \mathrm{e}^{-39{,}5t} - 0{,}61\cdot \mathrm{e}^{-0{,}5t}\ \text{A}$$

3.7.10 Overdamped Second Order Circuit Without Supply

The circuit represented in Fig. 3.42 is in steady state when, at $t = 0$, the switch opens. Calculate $U(s)$, transform of $u(t)$. Then, obtain the time expression, $u(t)$, for $t > 0$. What type of damping does the circuit have?

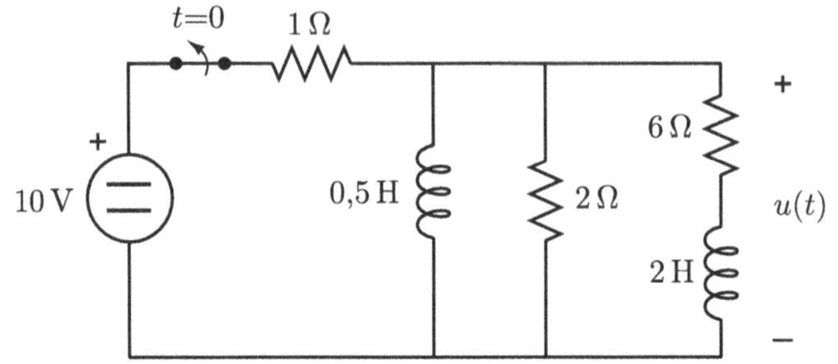

Fig. 3.42 Circuit considered in the problem under study

Solution. First, the inductor currents must be calculated before the switch opens, that is, $i_{L1}(0^-)$ and $i_{L2}(0^-)$. For this purpose, the dc steady-state circuit shown in Fig. 3.43 will be solved, where each inductor has been replaced by a short circuit.

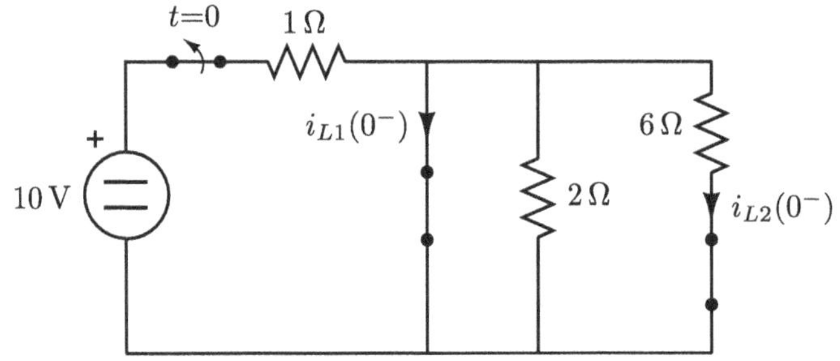

Fig. 3.43 Circuit in steady-state dc regime at $t = 0^-$

From the circuit in Fig. 3.43:

$$i_{L1}(0^-) = \frac{10}{1} = 10\,\text{A} \;\; ; \;\; i_{L2}(0^-) = 0\,\text{A}$$

The circuit for $t > 0$ is that shown in Fig. 3.44 in the Laplace domain. Applying the node analysis method, the following result is obtained:

$$\left(\frac{1}{0{,}5s} + \frac{1}{2} + \frac{1}{(2s+6)}\right) \cdot U(s) = \frac{-10}{s}$$

Rearranging terms:

$$U(s) = \frac{-10 \cdot (2s+6)}{s^2 + 8s + 12} = \frac{-10 \cdot (2s+6)}{(s+6)(s+2)}$$

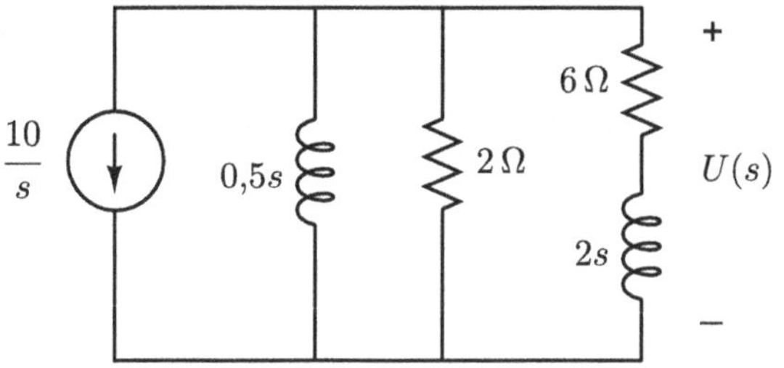

Fig. 3.44 Circuit in the Laplace domain

Analyzing the denominator of $U(s)$, it can be observed that it has two different real roots, corresponding to the natural response. Given the lack of a forced response, it is an overdamped second-order circuit.

Using the partial fraction expansion of $U(s)$ results in:

$$U(s) = \frac{-10 \cdot (2s + 6)}{(s + 6)(s + 2)} = \frac{A}{(s + 6)} + \frac{B}{(s + 2)}$$

The coefficients A and B are calculated as follows:

$$A = \left.\frac{-10 \cdot (2s + 6)}{(s + 2)}\right|_{s=-6} = \frac{-10 \cdot (2(-6) + 6)}{(-6 + 2)} = -15$$

$$B = \left.\frac{-10 \cdot (2s + 6)}{(s + 6)}\right|_{s=-2} = \frac{-10 \cdot (2(-2) + 6)}{(-2 + 6)} = -5$$

So:

$$U(s) = \frac{-15}{(s + 6)} + \frac{-5}{(s + 2)} \text{ V}$$

Finally, the inverse Laplace transform is taken:

$$u(t) = \mathscr{L}^{-1}[U(s)] = -15 \cdot e^{-6t} - 5 \cdot e^{-2t} \text{ V}$$

In $u(t)$ it can also be identified, based on the natural response, that it is an overdamped second order circuit.

3.7.11 Capacitors in Parallel. Impulse Response

The circuit represented in Fig. 3.45 is in steady state when, at $t = 0$, the switches change their position. Calculate the capacitor voltages and the current $i(t)$ for $t > 0$.

Solution. First, the voltage of capacitor will be obtained before the switches change their position ($t = 0^-$). For this purpose, it will be considered that the circuit is

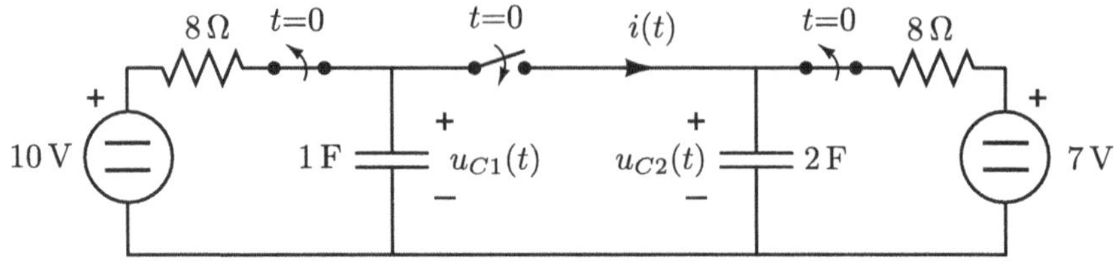

Fig. 3.45 Circuit considered in the problem under study

in steady-state before this moment. Since, in this situation, only dc-type excitation sources are present, the dc circuit shown in Fig. 3.46 will be solved, where each capacitor has been replaced by an open circuit.

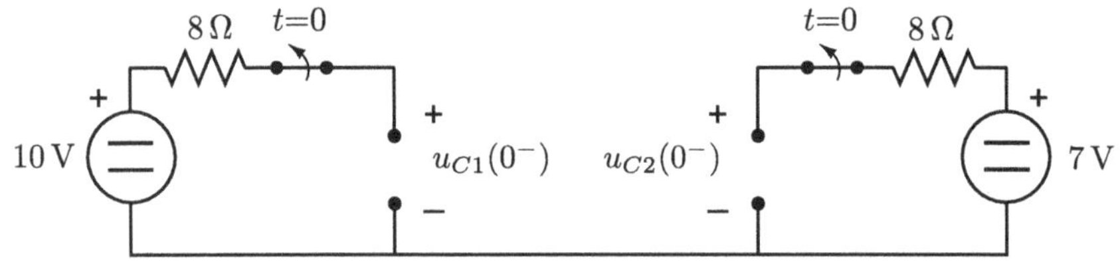

Fig. 3.46 Circuit at $t = 0^-$

From the circuit in Fig. 3.46:

$$u_{C1}(0^-) = 10\,\text{V} \;\; ; \;\; u_{C2}(0^-) = 7\,\text{V}$$

The circuit for $t > 0$ is that shown in Fig. 3.47 in the Laplace domain. Applying the node analysis method, the following result is obtained:

$$(s + 2s) \cdot U(s) = 10 + 14$$

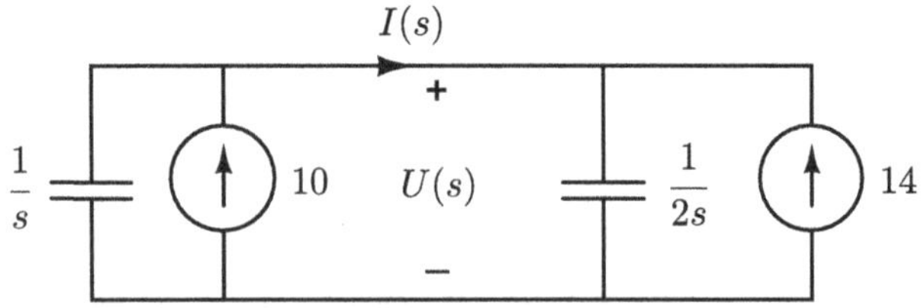

Fig. 3.47 Circuit in the Laplace domain

Rearranging terms:

$$U(s) = \frac{10 + 14}{s + 2s} = \frac{24}{3s} = \frac{8}{s}\,\text{V}$$

On the other hand:

$$I(s) = 10 - \frac{U(s)}{1/s} = 10 - \frac{8/s}{1/s} = 10 - 8 = 2\,\text{A}$$

Finally, the inverse Laplace transform is taken:

$$u(t) = \mathscr{L}^{-1}[U(s)] = 8\,\text{V}$$
$$i(t) = \mathscr{L}^{-1}[I(s)] = 2 \cdot \delta(t)\,\text{A}$$

3.7.12 First Order Under Circuit AC Supply

In the circuit represented in Fig. 3.48 it is known that $e(t) = 12\cos(t)$ V. Calculate $U(s)$, transform of $u(t)$. Then, determine the steady-state value and the time constant of the circuit.

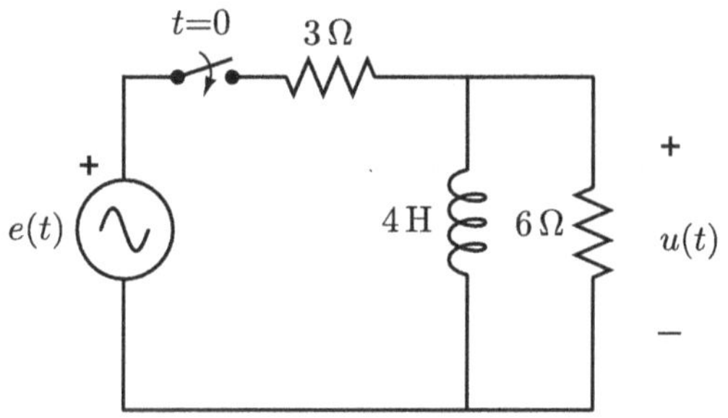

Fig. 3.48 Circuit considered in the problem under study

Solution. With the switch opened, the inductor current is zero, that is, $i_L(0^-) = 0\,\text{A}$.
The circuit for $t > 0$ is that shown in Fig. 3.49 in the Laplace domain, where:

$$E(s) = \mathscr{L}[e(t)] = \mathscr{L}[12\cos(t)] = \frac{12s}{s^2+1}$$

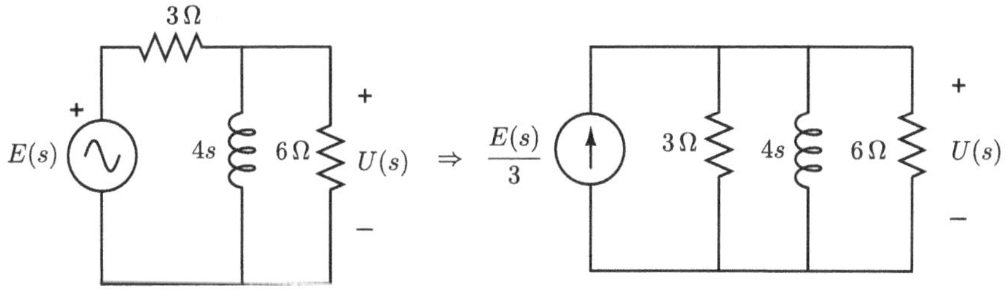

Fig. 3.49 Circuit in the Laplace domain

From the circuit in Fig. 3.49, the node analysis method can be applied with the following result:

$$\left[\frac{1}{3}+\frac{1}{4s}+\frac{1}{6}\right]\cdot U(s)=\frac{E(s)}{3}$$

Rearranging terms:

$$U(s)=\frac{E(s)/3}{\dfrac{6+4s+8s}{24s}}=\frac{\dfrac{1}{3}\cdot\dfrac{12s}{s^2+1}}{\dfrac{6+4s+8s}{24s}}=\frac{\dfrac{4s}{s^2+1}}{\dfrac{12s+6}{24s}}=\frac{8s^2}{(s^2+1)(s+0{,}5)}$$

Using the partial fraction expansion results in:

$$U(s)=\frac{8s^2}{(s-j)(s+j)(s+0{,}5)}=\frac{\overline{A}}{(s-j)}+\frac{\overline{A}^*}{(s+j)}+\frac{B}{(s+0{,}5)}$$

It can be noticed that the first two fractions correspond to the steady-state response, since their poles are associated with the Laplace transform of the ac term, while the third fraction corresponds to the natural response. This way, the time constant is calculated as:

$$\tau=\frac{1}{0{,}5}=2\,\text{s}$$

The coefficient $\overline{A}$ is obtained as follows:

$$\overline{A}=\left.\frac{8s^2}{(s+j)(s+0{,}5)}\right|_{s=j}=\frac{8\cdot(j)^2}{(j+j)(j+0{,}5)}\approx 3{,}58\angle 26{,}57^\circ$$

Finally, the steady-state response is determined with the inverse Laplace transform:

$$u^p(t)=\mathscr{L}^{-1}\left[\frac{3{,}58\angle 26{,}57^\circ}{(s-j)}+\frac{3{,}58\angle-26{,}57^\circ}{(s+j)}\right]=2\cdot 3{,}58\cdot\cos(t+26{,}57^\circ)\,\text{V}$$

3.7.13 First Order Circuit Under AC Supply

The circuit represented in Fig. 3.50 is in steady state when, at $t=0$, the switches change their position. Given that $e(t)=200\sin(100t+\varphi)$ V, with $0^\circ<\varphi<180^\circ$, calculate the value that φ should have, so that the circuit goes directly to a sinusoidal steady state after the switches change their position.

Solution. First, the inductor current will be obtained before the switches change their position ($t=0^-$). For this purpose, it will be considered that, at that moment, the circuit is in steady-state regime. Since, in this situation, only a dc-type source of excitation is present, the dc circuit shown in Fig. 3.51 will be solved, where the inductor has been replaced by a short circuit.

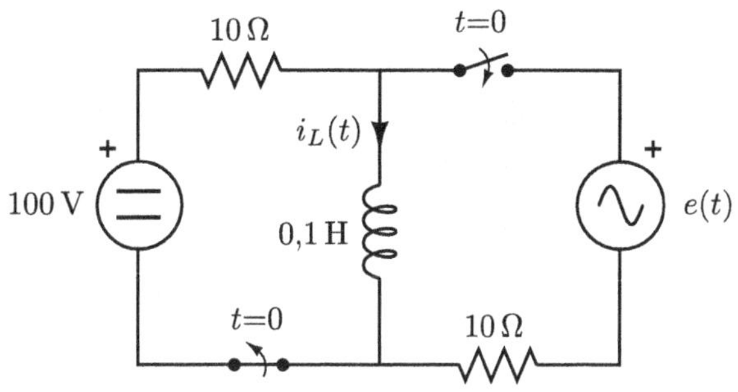

Fig. 3.50 Circuit considered in the problem under study

From the circuit in Fig. 3.51:

$$i_L(0^-) = \frac{100}{10} = 10\,\text{A}$$

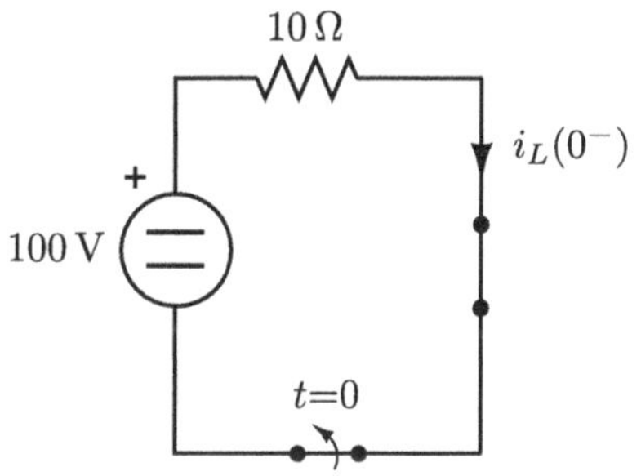

Fig. 3.51 Circuit at $t = 0^-$

The circuit for $t > 0$ is that shown in Fig. 3.52 in the Laplace domain, where:

$$E(s) = \mathscr{L}\,[e(t)] = \mathscr{L}\,[200\sin(100t + \varphi)] = \frac{200 \cdot (s \cdot \sin\varphi + 100 \cdot \cos\varphi)}{s^2 + 100^2}$$

From the circuit in Fig. 3.52:

$$\begin{aligned} I_L(s) &= \frac{E(s) + 1}{0,1s + 10} = \frac{\dfrac{200 \cdot (s \cdot \sin\varphi + 100 \cdot \cos\varphi)}{s^2 + 100^2} + 1}{0,1s + 10} \\ &= \frac{200 \cdot (s \cdot \sin\varphi + 100 \cdot \cos\varphi) + s^2 + 100^2}{0,1(s^2 + 100^2)(s + 100)} \end{aligned}$$

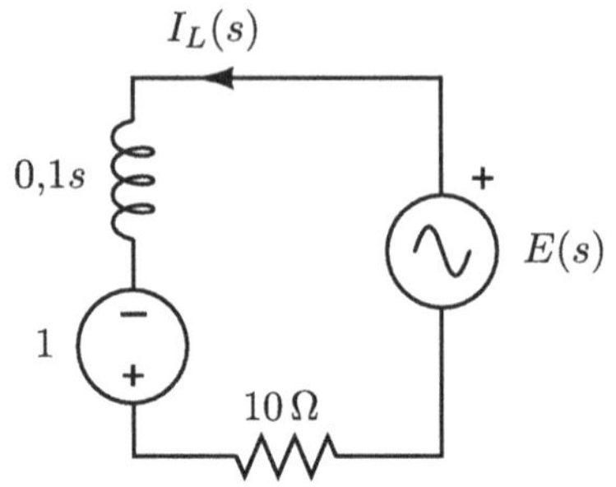

Fig. 3.52 Circuit in the Laplace domain

Using the partial fraction expansion results in:

$$I_L(s) = \frac{200 \cdot (s \cdot \sin\varphi + 100 \cdot \cos\varphi) + s^2 + 100^2}{0,1(s - 100j)(s + 100j)(s + 100)}$$
$$= \frac{\overline{A}}{(s - 100j)} + \frac{\overline{A}^*}{(s + 100j)} + \frac{B}{(s + 100)}$$

If the circuit goes directly to the steady-state regime, its natural response must be null, which implies that $B = 0$. This coefficient is calculated as follows:

$$B = \left.\frac{200 \cdot (s \cdot \sin\varphi + 100 \cdot \cos\varphi) + s^2 + 100^2}{0,1(s - 100j)(s + 100j)}\right|_{s=-100}$$
$$= \frac{200 \cdot ((-100) \cdot \sin\varphi + 100 \cdot \cos\varphi) + (-100)^2 + 100^2}{0,1(-100 - 100j)(-100 + 100j)}$$

Making $B = 0$, the following result is obtained:

$$200 \cdot ((-100) \cdot \sin\varphi + 100 \cdot \cos\varphi) + (-100)^2 + 100^2 = 0$$

Simplifying:

$$\cos\varphi - \sin\varphi = -1$$

Considering that

$$\cos^2\varphi + \sin^2\varphi = 1$$

so

$$(-1 + \sin\varphi)^2 + \sin^2\varphi = 1$$
$$\Rightarrow 1 - 2\sin\varphi + \sin^2\varphi + \sin^2\varphi = 1$$
$$\Rightarrow \sin\varphi \cdot (\sin\varphi - 1) = 0$$

whose possible solutions are:

$$\varphi = 0^\circ \ ; \ \varphi = 180^\circ \ ; \ \varphi = 90^\circ$$

According to the statement, $0^\circ < \varphi < 180^\circ$, therefore the valid solution is:

$$\varphi = 90^\circ$$

3.7.14 First Order Circuit Under Dependent Source and DC Supply

The circuit represented in Fig. 3.53 is in steady state before the switch changes its position. Determine $U(s)$ and $U_1(s)$, transformed of $u(t)$ and $u_1(t)$ respectively. Then, obtain the time expressions, $u(t)$ and $u_1(t)$, for $t > 0$.

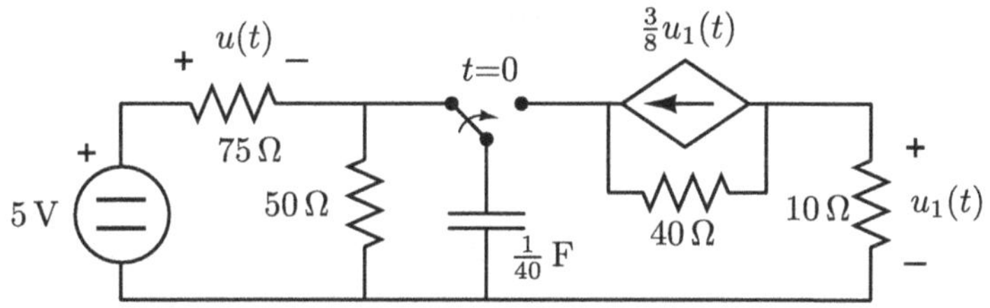

Fig. 3.53 Circuit considered in the problem under study

Solution. First, the capacitor voltage will be obtained before the switch changes its position, that is, at instant $t = 0^-$. For this purpose, it will be considered that the circuit is in steady-state regime at that moment. Since only dc-type excitation sources are present, the dc circuit shown in Fig. 3.54 will be solved, where the capacitor has been replaced by an open circuit.

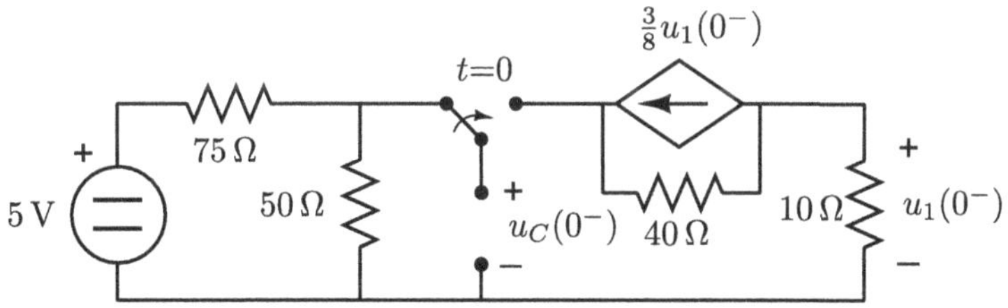

Fig. 3.54 Circuit at $t = 0^-$

From Fig. 3.54 the capacitor voltage at $t = 0^-$ is obtained:

$$u_C(0^-) = 5 \cdot \frac{50}{50 + 75} = 2\,\text{V}$$

The circuit for $t > 0$ is that shown in Fig. 3.55 in the Laplace domain.

From the circuit in Fig. 3.55:

$$U(s) = \frac{5}{s} \cdot \frac{75}{75 + 50} = \frac{3}{s}$$

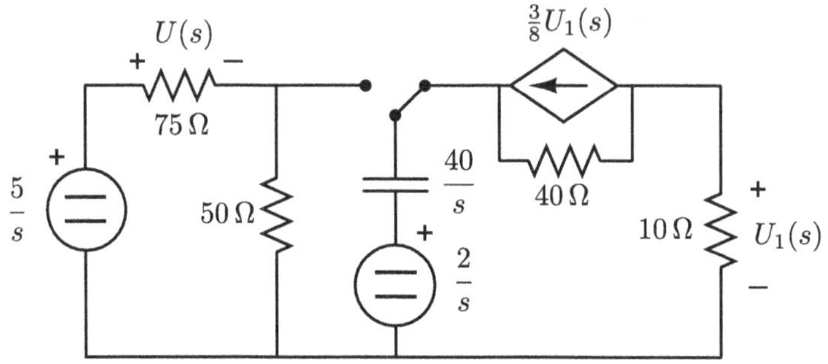

Fig. 3.55 Circuit in the Laplace domain

The right side of the circuit in Fig. 3.55 will be simplified by substituting the dependent current source in parallel with the resistor for a dependent voltage source in series with the resistor, as shown in Fig. 3.56.

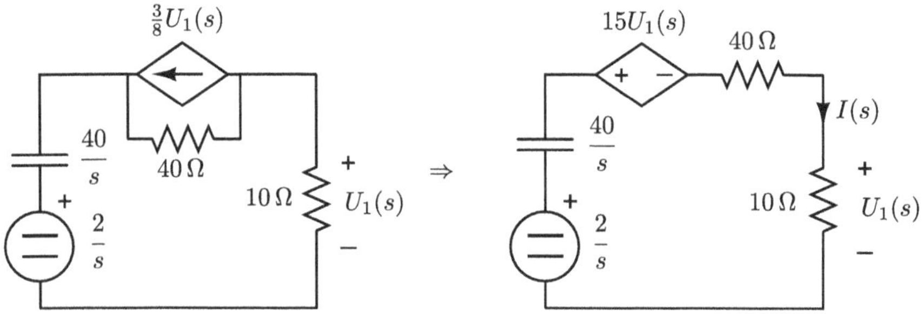

Fig. 3.56 Circuit simplification in the Laplace domain

From the circuit in Fig. 3.56:

$$\frac{2}{s} - 15U_1(s) = \left(\frac{40}{s} + 40 + 10\right) \cdot I(s)$$

$$U_1(s) = 10 \cdot I(s)$$

Solving the previous equation, $U_1(s)$ is obtained:

$$U_1(s) = \frac{0{,}1}{s + \frac{1}{5}}$$

Finally, the inverse Laplace transform is taken:

$$u(t) = \mathscr{L}^{-1}\left[U(s)\right] = 3\,\mathrm{V}$$
$$u_1(t) = \mathscr{L}^{-1}\left[U_1(s)\right] = 0{,}1 \cdot \mathrm{e}^{-t/5}\,\mathrm{V}$$

3.7.15 Magnetically Coupled Coils

The circuit represented in Fig. 3.57 is in steady-state when the switch is closed at $t = 0$. Calculate $I(s)$, transform of $i(t)$. Then, deduce the order of the circuit and, depending on this, obtain the time constant or the damping coefficient and the resonant frequency. Finally, calculate $i(t)$ for $t > 0$.

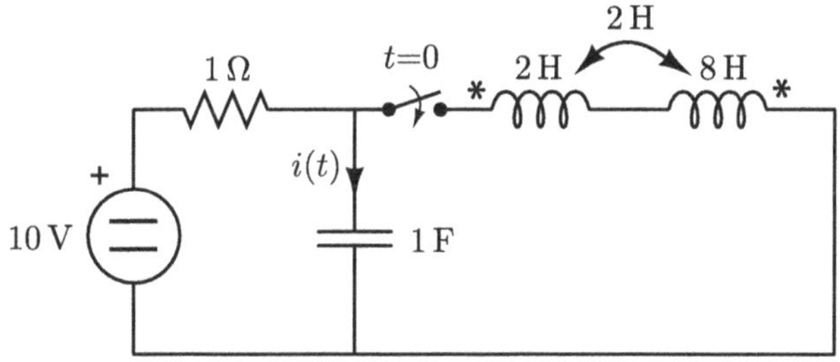

Fig. 3.57 Circuit considered in the problem under study

Solution. Before the switch closes, the current through the inductors is zero. On the other hand, to calculate the capacitor voltage before the switch closes, the dc circuit shown in Fig. 3.58 will be solved.

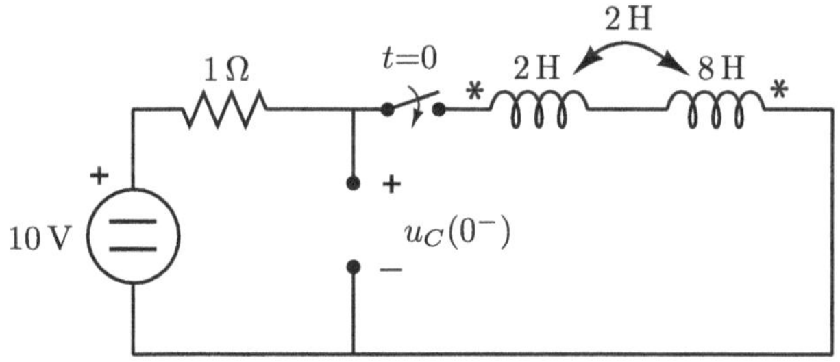

Fig. 3.58 Circuit at $t = 0^-$

According to Fig. 3.58:

$$u_C(0^-) = 10\,\mathrm{V}$$

The circuit for $t > 0$ is that shown in Fig. 3.59 in the Laplace domain, which can be simplified by associating series impedances and dependent sources, obtaining the circuit represented in Fig. 3.60, where the mesh currents have been represented, since this method will be used for the circuit resolution.

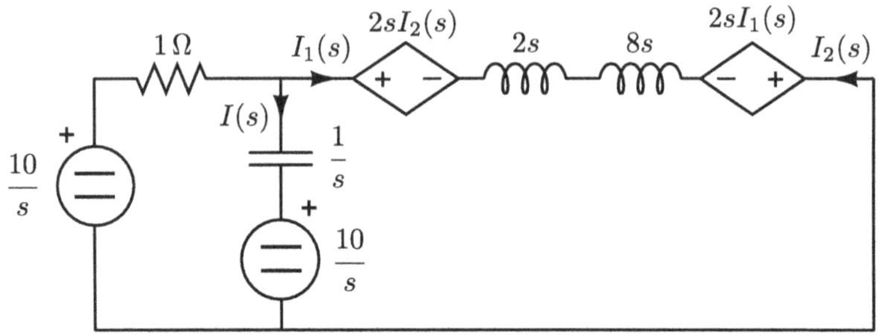

Fig. 3.59 Circuit in the Laplace domain

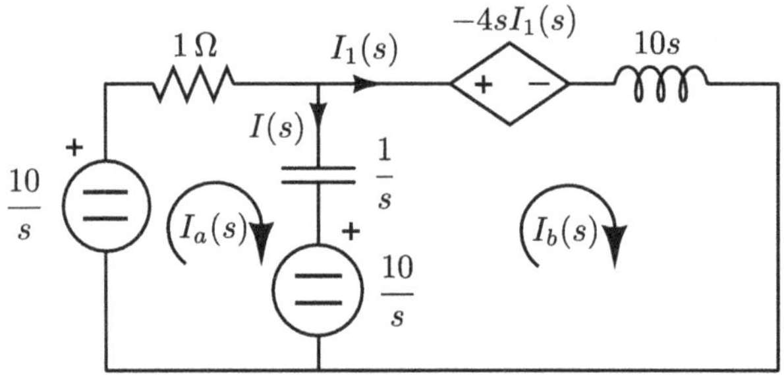

Fig. 3.60 Simplified circuit in the Laplace domain

From the circuit in Fig. 3.60, applying the mesh analysis results in:

$$\begin{bmatrix} 1+\frac{1}{s} & \frac{-1}{s} \\ \frac{-1}{s} & \frac{1}{s}+10s \end{bmatrix} \begin{bmatrix} I_a(s) \\ I_b(s) \end{bmatrix} = \begin{bmatrix} \frac{10}{s}-\frac{10}{s} \\ \frac{10}{s}+4sI_1(s) \end{bmatrix}$$

Considering that $I_1(s) = I_b(s)$:

$$\begin{bmatrix} 1+\frac{1}{s} & \frac{-1}{s} \\ \frac{-1}{s} & \frac{1}{s}+10s \end{bmatrix} \begin{bmatrix} I_a(s) \\ I_b(s) \end{bmatrix} = \begin{bmatrix} \frac{10}{s}-\frac{10}{s} \\ \frac{10}{s}+4sI_b(s) \end{bmatrix}$$

Rearranging terms and simplifying, the mesh equations are finally obtained:

$$\begin{bmatrix} 1+\dfrac{1}{s} & \dfrac{-1}{s} \\ \dfrac{-1}{s} & \dfrac{1}{s}+6s \end{bmatrix} \begin{bmatrix} I_a(s) \\ I_b(s) \end{bmatrix} = \begin{bmatrix} 0 \\ \dfrac{10}{s} \end{bmatrix}$$

This system of equations can be simplified by multiplying all the terms by s, resulting in:

$$\begin{bmatrix} 1+s & -1 \\ -1 & 1+6s^2 \end{bmatrix} \begin{bmatrix} I_a(s) \\ I_b(s) \end{bmatrix} = \begin{bmatrix} 0 \\ 10 \end{bmatrix}$$

Solving the previous equations, the mesh currents are obtained:

$$I_a(s) = \frac{\frac{10}{6}}{s(s^2+s+\frac{1}{6})} \quad ; \quad I_b(s) = \frac{\frac{10}{6}\cdot(1+s)}{s(s^2+s+\frac{1}{6})}$$

Therefore, the expression of the current $I(s)$ is:

$$I(s) = I_a(s) - I_b(s) = \frac{\frac{10}{6}}{s(s^2+s+\frac{1}{6})} - \frac{\frac{10}{6}\cdot(1+s)}{s(s^2+s+\frac{1}{6})} = \frac{-\frac{10}{6}}{s^2+s+\frac{1}{6}}$$

Considering the denominator of $I(s)$, it can be noticed that the circuit is second order. Furthermore, the damping coefficient and resonant frequency are calculated as follows:

$$2\alpha = 1 \;\Rightarrow\; \alpha = \frac{1}{2}\,\mathrm{s}^{-1}$$

$$\omega_0^2 = \frac{1}{6} \;\Rightarrow\; \omega_0 = \frac{1}{\sqrt{6}}\,\mathrm{rad\,s}^{-1}$$

Since $\alpha > \omega_0$, the natural response is overdamped.
Using the partial fraction expansion of $I(s)$ results in:

$$I(s) = \frac{-\frac{10}{6}}{s^2+s+\frac{1}{6}} = \frac{-\frac{10}{6}}{(s+0{,}211)(s+0{,}789)} = \frac{A}{(s+0{,}211)} + \frac{B}{(s+0{,}789)}$$

The coefficients A and B are calculated as follows:

$$A = \left.\frac{-\frac{10}{6}}{(s+0{,}789)}\right|_{s=-0{,}211} = \frac{-\frac{10}{6}}{(-0{,}211+0{,}789)} \approx -2{,}88$$

$$B = \left.\frac{-\frac{10}{6}}{(s+0{,}211)}\right|_{s=-0{,}789} = \frac{-\frac{10}{6}}{(-0{,}789+0{,}211)} \approx 2{,}88$$

So:

$$I(s) = \frac{-2{,}88}{(s+0{,}211)} + \frac{2{,}88}{(s+0{,}789)} \text{ A}$$

Finally, the inverse Laplace transform is taken:

$$i(t) = \mathscr{L}^{-1}[I(s)] = -2{,}88 \cdot e^{-0{,}211t} + 2{,}88 \cdot e^{-0{,}789t} \text{ A}$$

3.7.16 Magnetically Coupled Coils

The circuit represented in Fig. 3.61 is in steady state. At $t = 0$, the switch changes its position. Determine $I(s)$, transform of $i(t)$. Using this result, determine the order of the circuit and, depending on this, the time constant or the damping coefficient and the resonant frequency.

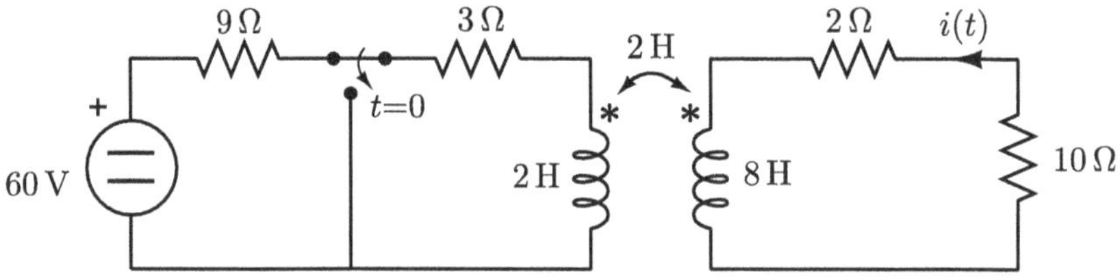

Fig. 3.61 Circuit considered in the problem under study

Solution. First, the current through each inductor will be calculated before the switch changes its position. For this purpose, it will be considered that the circuit is in steady-state regime before this change. Since only a dc-type source of excitation is present, the dc circuit shown in Fig. 3.62 will be solved, where the inductors have been replaced by short circuits.

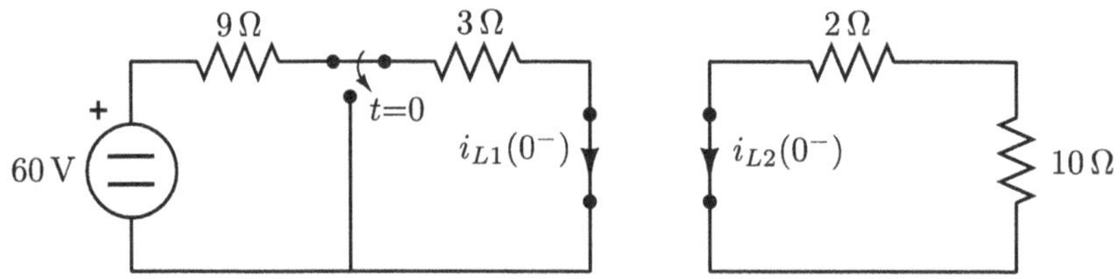

Fig. 3.62 Circuit at $t = 0^-$

From the circuit in Fig. 3.62:

$$i_{L1}(0^-) = \frac{60}{12} = 5 \text{ A} \ ; \ i_{L2}(0^-) = 0 \text{ A}$$

The circuit for $t > 0$ is that shown in Fig. 3.63 in the Laplace domain.

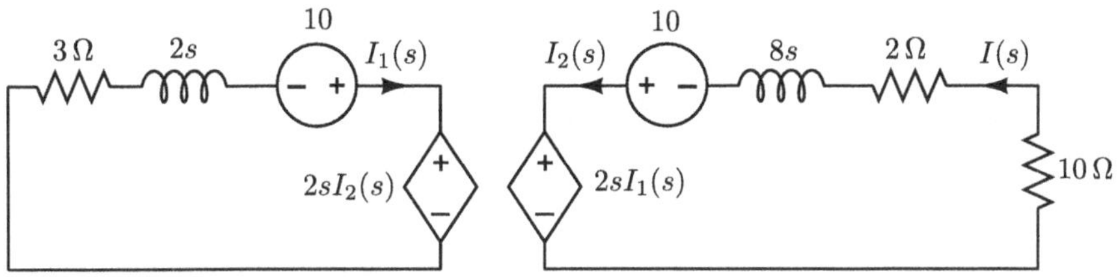

Fig. 3.63 Circuit in the Laplace domain

Applying Kirchhoff's voltage law in both sides of the circuit results in:

$$(3 + 2s) \cdot I_1(s) - 10 + 2sI_2(s) = 0$$
$$(2 + 10 + 8s) \cdot I_2(s) - 10 + 2sI_1(s) = 0$$

Solving the previous equations, $I_2(s)$ is obtained, which according to the circuit in Fig. 3.63, matches to $I(s)$:

$$I(s) = I_2(s) = \frac{30/12}{s^2 + 4s + 3}\,\text{A}$$

Considering the denominator of $I(s)$, it can be noticed that the circuit is second order. Furthermore, the damping coefficient and resonant frequency are calculated as follows:

$$2\alpha = 4 \;\Rightarrow\; \alpha = 2\,\text{s}^{-1}$$
$$\omega_0^2 = 3 \;\Rightarrow\; \omega_0 = \sqrt{3}\,\text{rad}\,\text{s}^{-1}$$

3.7.17 Second Order Circuit Under Impulse Supply

In the circuit represented in Fig. 3.64, it is known that $u_C(0^-) = 0\,\text{V}$ and $i_L(0^-) = 0\,\text{A}$. Calculate $I(s)$, transform of $i(t)$. Then, obtain the corresponding time expression, $i(t)$, for $t > 0$.

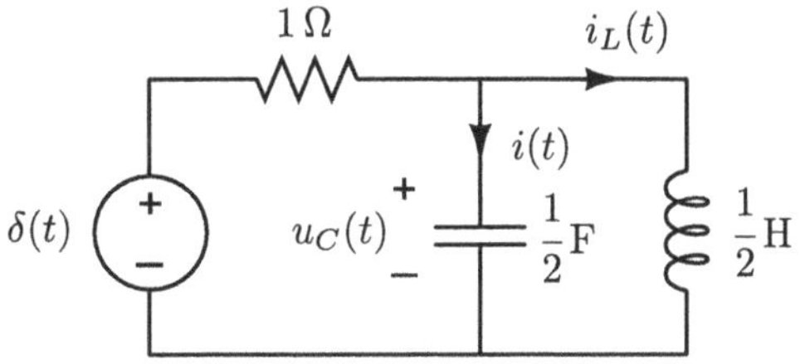

Fig. 3.64 Circuit considered in the problem under study

Solution. The circuit for $t > 0$ is that shown in Fig. 3.65 in the Laplace domain.

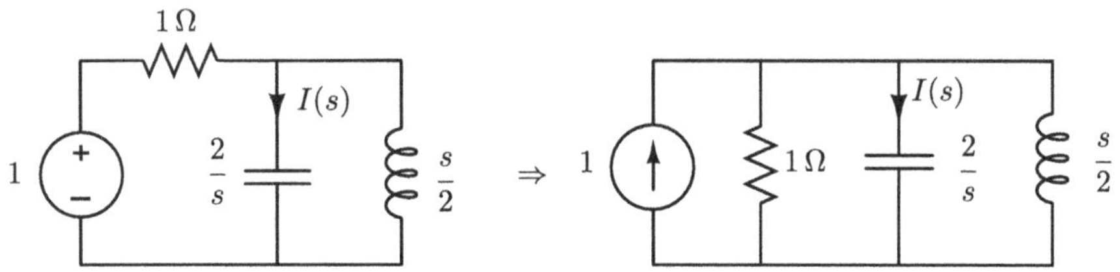

Fig. 3.65 Circuit in the Laplace domain

From the circuit in Fig. 3.65, the current $I(s)$ is obtained by applying the concept of current divider:

$$I(s) = 1 \cdot \frac{\frac{s}{2}}{\frac{s}{2} + 1 + \frac{2}{s}} = \frac{s^2}{s^2 + 2s + 4}$$

It can be noticed that both numerator and denominator of $I(s)$ have the same order, therefore, to decompose into simple fractions, the division of polynomials shown in Fig. 3.66 must be first applied.

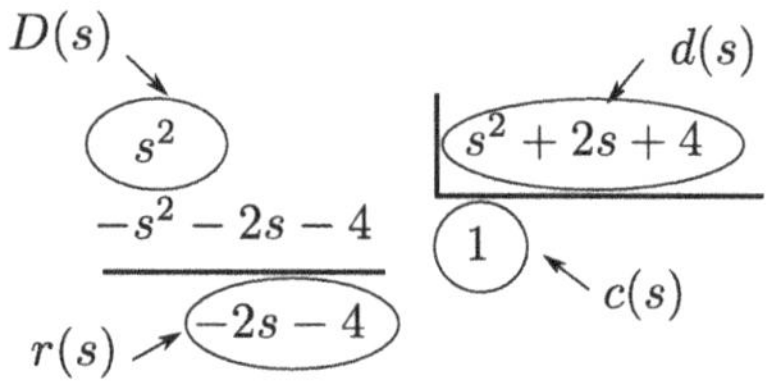

Fig. 3.66 Polynomial division

Considering the scheme in Fig. 3.66, $I(s)$ can be expressed as follows:

$$I(s) = \frac{D(s)}{d(s)} = c(s) + \frac{r(s)}{d(s)} = 1 + \frac{-2s - 4}{s^2 + 2s + 4} = 1 + \frac{-2s - 4}{(s + 1 - \sqrt{3}j)(s + 1 + \sqrt{3}j)}$$

Using the partial fraction expansion results in:

$$I(s) = 1 + \frac{-2s - 4}{(s + 1 - \sqrt{3}j)(s + 1 + \sqrt{3}j)} = 1 + \frac{\overline{A}}{(s + 1 - \sqrt{3}j)} + \frac{\overline{A}^*}{(s + 1 + \sqrt{3}j)}$$

The coefficient $\overline{A}$ is calculated as follows:

$$\overline{A} = \left.\frac{-2s - 4}{(s + 1 + \sqrt{3}j)}\right|_{s=-1+\sqrt{3}j} = \frac{-2 \cdot (-1 + \sqrt{3}j) - 4}{(-1 + \sqrt{3}j + 1 + \sqrt{3}j)} \approx 1{,}15\angle 150^\circ$$

So:

$$I(s) = 1 + \frac{1{,}15\angle 150^\circ}{(s+1-\sqrt{3}j)} + \frac{1{,}15\angle -150^\circ}{(s+1+\sqrt{3}j)}\ \mathrm{A}$$

Finally, the inverse Laplace transform is taken:

$$i(t) = \mathscr{L}^{-1}\left[I(s)\right] = \delta(t) + 2 \cdot 1{,}15 \cdot \mathrm{e}^{-t}\cos(\sqrt{3}t + 150^\circ)\ \mathrm{A}$$

3.7.18 Second Order Circuit Under Exponential Supply

The circuit represented in Fig. 3.67 is in steady state. At t=0, the switch changes its position. Calculate $U_C(s)$, transform of $u_C(t)$. From this result, determine the type of damping as well as the damping coefficient and the resonant frequency. Finally, calculate the time expression, $u_C(t)$, for $t > 0$.

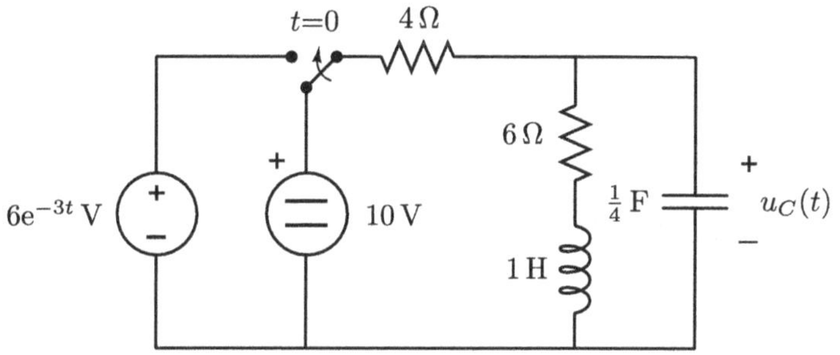

Fig. 3.67 Circuit considered in the problem under study

Solution. First, the inductor current and the capacitor voltage will be obtained before the switch changes its position, that is, at $t = 0^-$. For this purpose, it will be considered that the circuit is in steady-state regime at that moment. Since only dc-type sources of excitation are present, the dc circuit shown in Fig. 3.68 will be solved.

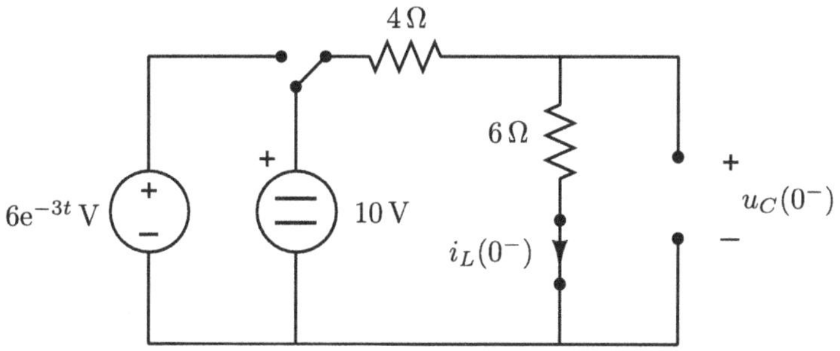

Fig. 3.68 Circuit at $t = 0^-$

From the circuit in Fig. 3.68:

$$i_L(0^-) = \frac{10}{6+4} = 1\,\mathrm{A}$$
$$u_C(0^-) = 6 \cdot 1 = 6\,\mathrm{V}$$

The circuit for $t > 0$ is that shown in Fig. 3.69 in the Laplace domain.

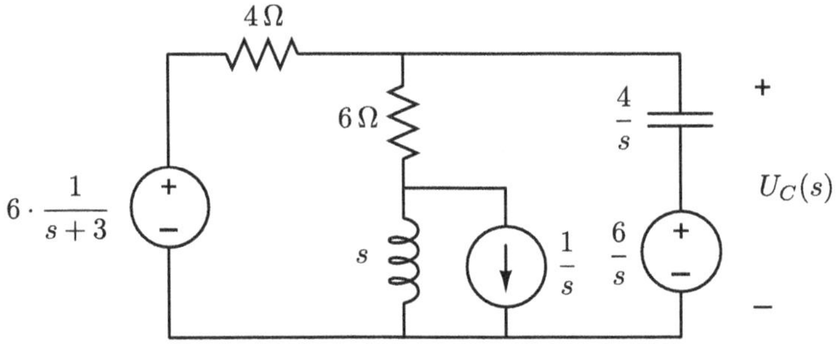

Fig. 3.69 Circuit in the Laplace domain

The circuit depicted in Fig. 3.69 can be simplified by source conversion, leading to the circuit in Fig. 3.70.

From the circuit in Fig. 3.70, applying the node analysis method results in:

$$\left(\frac{1}{4} + \frac{1}{6+s} + \frac{s}{4}\right) \cdot U_C(s) = \frac{6/4}{s+3} + \frac{6}{4} - \frac{1}{s+6}$$

Solving the previous equation, $U_C(s)$ is obtained:

$$U_C(s) = \frac{6s^2 + 56s + 132}{(s+3) \cdot \left(s^2 + 7s + 10\right)}$$

Considering the denominator of $U_C(s)$, it can be noticed that the factor $(s+3)$ corresponds to the excitation source while the factor $(s^2 + 7s + 10)$ is the one related to the natural response. Therefore, the circuit is second order and the damping coefficient and the resonant frequency are calculated as follows:

$$2\alpha = 7 \;\Rightarrow\; \alpha = 3{,}5\,\mathrm{s}^{-1}$$
$$\omega_0^2 = 10 \;\Rightarrow\; \omega_0 \approx 3{,}16\,\mathrm{rad\,s}^{-1}$$

Since $\alpha > \omega_0$, the natural response is overdamped.

Using the partial fraction expansion of $U_C(s)$ results in:

$$U_C(s) = \frac{6s^2 + 56s + 132}{(s+3)(s+5)(s+2)} = \frac{A}{(s+3)} + \frac{B}{(s+5)} + \frac{C}{(s+2)}$$

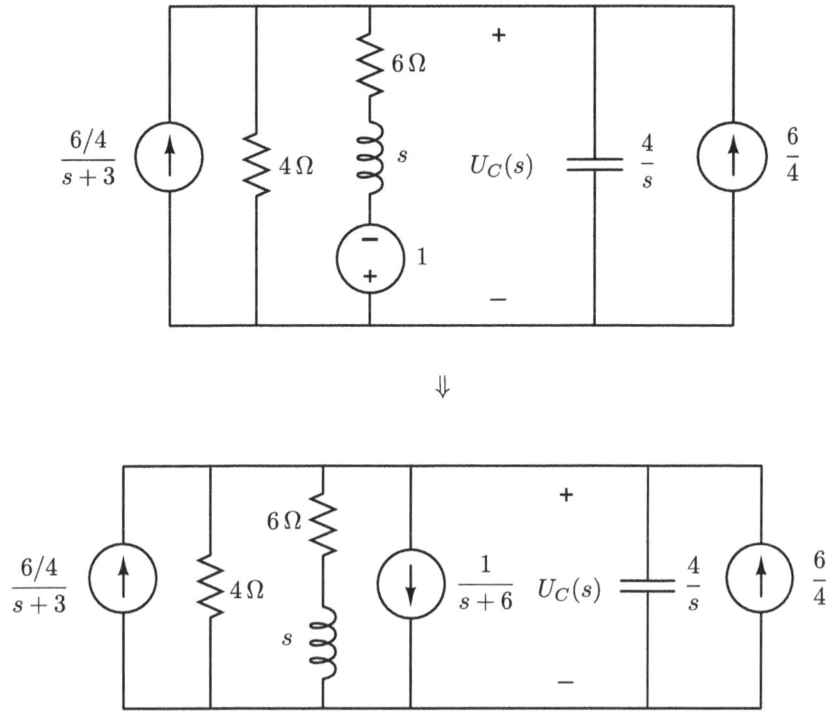

Fig. 3.70 Simplified circuit in the Laplace domain

The coefficients A, B and C are calculated as follows:

$$A = \left.\frac{6s^2 + 56s + 132}{(s+5)(s+2)}\right|_{s=-3} = \frac{6(-3)^2 + 56(-3) + 132}{(-3+5)(-3+2)} = -9$$

$$B = \left.\frac{6s^2 + 56s + 132}{(s+3)(s+2)}\right|_{s=-5} = \frac{6(-5)^2 + 56(-5) + 132}{(-5+3)(-5+2)} \approx 0{,}33$$

$$C = \left.\frac{6s^2 + 56s + 132}{(s+3)(s+5)}\right|_{s=-2} = \frac{6(-2)^2 + 56(-2) + 132}{(-2+3)(-2+5)} \approx 14{,}67$$

So:

$$U_C(s) = \frac{-9}{(s+3)} + \frac{0{,}33}{(s+5)} + \frac{14{,}67}{(s+2)}\ \mathrm{V}$$

Finally, the inverse Laplace transform is taken:

$$u_C(t) = \mathscr{L}^{-1}\left[U_C(s)\right] = -9 \cdot \mathrm{e}^{-3t} + 0{,}33 \cdot \mathrm{e}^{-5t} + 14{,}67 \cdot \mathrm{e}^{-2t}\ \mathrm{V}$$

Analyzing $u_C(t)$ it can be remarked that the first term corresponds to the forced response of the exponential excitation source, while the remaining two terms correspond to the natural overdamped response.

3.7.19 Impulse Response

The circuit represented in Fig. 3.71 is in steady state. At $t = 0$, the switch changes its position. Calculate $U(s)$, transform of $u(t)$. Then, obtain the time expression, $u(t)$, for $t > 0$.

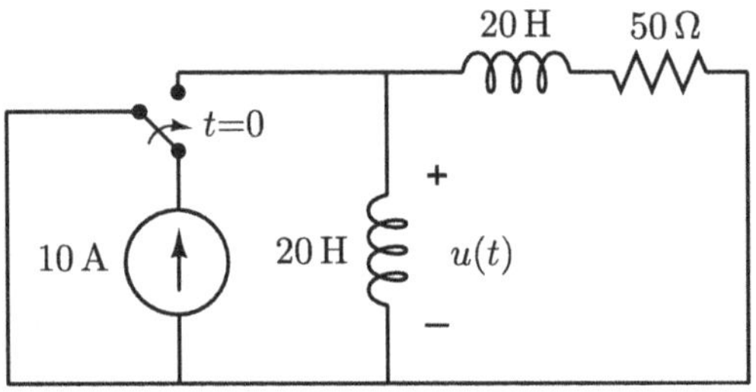

Fig. 3.71 Circuit considered in the problem under study

Solution. Before the switch changes its position, the circuit is in state-state regime and it can be verified that the current through both inductors is zero.

The circuit for $t > 0$ is that shown in Fig. 3.72 in the Laplace domain.

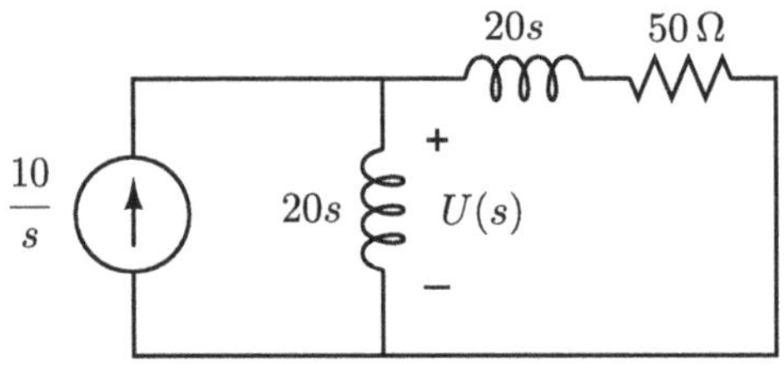

Fig. 3.72 Circuit in the Laplace domain

From the circuit in Fig. 3.72, the equivalent impedance is obtained as follows:

$$Z_{eq}(s) = \frac{20s \cdot (20s + 50)}{20s + 20s + 50} = \frac{10s \cdot (s + \frac{5}{2})}{s + \frac{5}{4}}$$

Therefore, the voltage $U(s)$ is calculated as:

$$U(s) = \frac{10}{s} \cdot Z_{eq}(s) = \frac{10}{s} \cdot \frac{10s \cdot (s + \frac{5}{2})}{s + \frac{5}{4}} = \frac{100 \cdot (s + \frac{5}{2})}{s + \frac{5}{4}}$$

It can be noticed that both numerator and denominator of $U(s)$ have the same order, therefore, to decompose the expression into simple fractions, it is necessary to previously make the division of polynomials shown in Fig. 3.73.

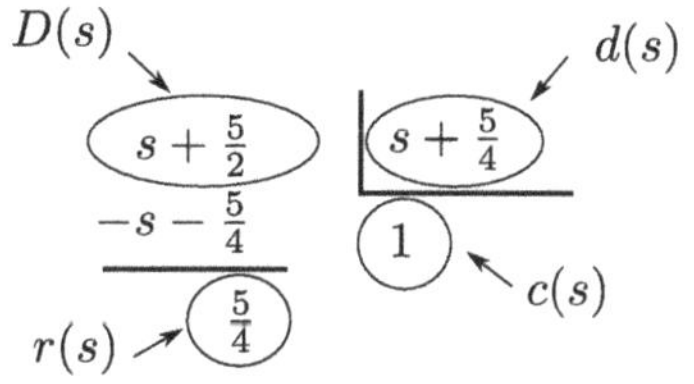

Fig. 3.73 Polynomial quotient

Considering Fig. 3.73, $U(s)$ can be expressed as follows:

$$U(s) = 100 \cdot \left[\frac{D(s)}{d(s)}\right] = 100 \cdot \left[c(s) + \frac{r(s)}{d(s)}\right] = 100 \cdot \left[1 + \frac{5/4}{s + \frac{5}{4}}\right] = 100 + \frac{125}{s + \frac{5}{4}} \text{ V}$$

Finally, the inverse Laplace transform is taken:

$$u(t) = \mathscr{L}^{-1}[U(s)] = 100 \cdot \delta(t) + 125 \cdot e^{-5t/4} \text{ V}$$

The manufacturer's authorised representative in the EU is Springer Nature Customer Service Centre GmbH, Europaplatz 3, 69115 Heidelberg, Germany. If you have any concerns regarding our products, please contact ProductSafety@springernature.com

Printed and bound by CPI Group (UK) Ltd, Croydon, CR0 4YY
30/09/2026
02225690-0005